GLOBAL ENERGY STRATEGIES

Living with Restricted Greenhouse Gas Emissions

ENVIRONMENTAL SCIENCE RESEARCH

Recent Volumes in this Series

Volume 39 — GENETIC TOXICOLOGY OF COMPLEX MIXTURES
Edited by Michael D. Waters, F. Bernard Daniel, Joellen Lewtas, Martha M. Moore, and Stephen Nesnow

Volume 40 — NITROARENES: Occurrence, Metabolism, and Biological Impact
Edited by Paul C. Howard, Stephen S. Hecht, and Frederick A. Beland

Volume 41 — ENVIRONMENTAL BIOTECHNOLOGY FOR WASTE TREATMENT
Edited by Gary S. Sayler, Robert Fox, and James W. Blackburn

Volume 42 — CHEMISTRY FOR THE PROTECTION OF THE ENVIRONMENT
Edited by L. Pawlowski, W. J. Lacy, and J. J. Dlugosz

Volume 43 — PRIMARY PRODUCTIVITY AND BIOGEOCHEMICAL CYCLES IN THE SEA
Edited by Paul G. Falkowski and Avril D. Woodhead

Volume 44 — SECONDARY-METABOLITE BIOSYNTHESIS AND METABOLISM
Edited by Richard J. Petroski and Susan P. McCormick

Volume 45 — GLOBAL CLIMATE CHANGE: Linking Energy, Environment, Economy, and Equity
Edited by James C. White

Volume 46 — PRODUCED WATER: Technological/Environmental Issues and Solutions
Edited by James P. Ray and F. Rainer Engelhardt

Volume 47 — GLOBAL ENERGY STRATEGIES: Living with Restricted Greenhouse Gas Emissions
Edited by James C. White

GLOBAL ENERGY STRATEGIES

Living with Restricted Greenhouse Gas Emissions

Edited by

James C. White

Cornell University
Ithaca, New York

Springer Science+Business Media, LLC

Library of Congress Cataloging-in-Publication Data

Global energy strategies : living with restricted greenhouse gas emissions / edited by James C. White.
p. cm. -- (Environmental science research ; v. 47)
"Proceedings of the Center for Environmental Information's Ninth International Conference on Global Energy Strategies: Living with Restricted Greenhouse Gas Emissions, held December 8-10, 1992, in Washington, D.C."--T.p. verso.
Includes bibliographical references and index.
ISBN 978-1-4899-1258-9 ISBN 978-1-4899-1256-5 (eBook)
DOI 10.1007/978-1-4899-1256-5
1. Greenhouse gases--Environmental aspects--Congresses.
2. Renewable energy resources--Congresses. 3. Energy policy--Congresses. I. White, James C. (James Carrick), 1916- .
II. Center for Environmental Information (U.S.) III. International Conference on Global Energy Strategies: Living with Restricted Greenhouse Gas Emissions (9th : 1992 : Washington, D.C.)
IV. Series.
TD885.5.G73G58 1994
363.73'87--dc20 94-14491
CIP

Proceedings of the Center for Environmental Information's Ninth International Conference on Global Energy Strategies: Living with Restricted Greenhouse Gas Emissions, held December 8–10, 1992, in Washington, D.C.

ISBN 978-1-4899-1258-9

Originally published by Plenum Press, New York in 1993
Softcover reprint of the hardcover 1st edition 1993

PREFACE

The world is getting warmer. Among scientists concerned with global climate change this is the broad consensus. How fast and by how much, are questions which cannot be answered quantitatively, but the probability of rising temperatures must be faced in a prudent manner - there is enough certainty of change so that we must anticipate and prepare before irreparable damage is done to our world. Even if it isn't going to be as bad as some people think, the actions we propose will benefit the earth and give us a kind of insurance.

The root of the change is population growth, and its attendant demand for energy. While the developed world expects to hold future emissions relatively steady, the developing countries, where population growth is most rampant, will expand the use of energy as they aspire to a better quality of life. If greater energy use is inevitable it behooves us to produce that energy in the least objectionable manner, and to produce it where the cost is lowest in dollars, GNP, and environmental change.

With present technology it seems impossible to meet rising energy needs without increasing the yearly output of greenhouse gases. Yet the Rio Conference Agreement, signed by more countries than any other agreement in history, seeks to hold emissions near the 1990 level. What is not obvious is that holding the emissions levels constant still results in increased accumulations in the atmosphere and the attendant deleterious effects. Without non-carbon dioxide produced energy we cannot hope to control atmospheric composition.

This book contains the papers presented at a Washington, D.C. meeting, December 1992. The program, established by an outstanding advisory committee, concentrates on the various potentials for controlling, reducing and ameliorating greenhouse gas emissions; on energy policies and strategies in the developed and the developing world; and on integrating the concerns of all nations.

The speakers on non-fossil fuel energy were charged to address their topics around a common set of questions; energy feasibility, costs in Kwh and GNP, time scale of availability, implications for UN and other global decisions, greenhouse gas impacts, barriers and incentives to use, societal implications, energy storage capability, and international competitiveness.

The reader will find the papers on U.S. state and federal policies, on the perceptions of other parts of the world and on integrating global concerns to be most useful. No-regrets policies are well documented, but other actions are also emphasized. Finally, some of the discussions between speakers and the audience are included.

The conference and this volume could not have come to pass without the exemplary contributions of Susanna Cordts who organized and managed the meeting. We also thank the School of Hotel Administration, Cornell University, for providing office facilities for preparation of the manuscript. A special thanks goes to Peggy Sipes without whose good humor, fantastic skill with the computer and infinite patience with the editor publication would have taken a lot longer.

This book contains the work of many leaders in all facets of the global climate change problem. Their ideas are novel and challenging and we trust that this volume will be useful to scholars in many disciplines.

James C. White
Cornell University

ABOUT THE CENTER FOR ENVIRONMENTAL INFORMATION

The Center for Environmental Information (CEI) was established in Rochester, New York, in 1974 as an answer to the growing dilemma of where to find timely, accurate and comprehensive information on environmental issues. To meet this need for current and comprehensive information, CEI has developed a multi-faceted program of publications, educational programs and information services. It is a private, nonprofit organization funded by membership dues, fees, contracts, grants and contributions.

The Center remains today a Rochester-based organization, but its services now reach far beyond the local community, reflecting the increasing number, scope and complexity of problems affecting the environment.

CEI acts as a catalyst to advance the public agenda toward soundly conceived environmental policies. CEI's communication network provides a link among the scientific community, educators, decision makers and the public, so that informed action follows the free interchange of information and ideas.

CONFERENCE STEERING COMMITTEE

Christopher Bernabo, President, Science and Policy Associates, Inc.
Alex Cristofaro, Director, Air and Energy Policy Division, U.S. Environmental Protection Agency
Roger C. Dower, Director, Climate, Energy and Pollution Program, World Resources Institute
Christopher Fox, Assistant Director, Center for Global Change, University of Maryland at College Park
Howard Gruenspecht, Associate Deputy Undersecretary for Program Analysis, U.S. Department of Energy
T.J. Glauthier, Director, Energy and Climate Change Program, World Wildlife Fund
Charles Herrick, Associate Director for Environmental Trends, Council on Environmental Quality
Gordon J. MacDonald, Institute on Global Conflict and Cooperation, University of California, San Diego
Lester Machta, Air Resources Laboratory, NOAA
Donna Mercado, Manager, Environmental Analysis, American Gas Association
Ralph Perhac, Former Senior Scientific Advisor, Environment Division, Electric Power Research Institute (EPRI)
Richard Richels, Director, Energy Analysis and Planning Department, EPRI
Elizabeth Thorndike, President, Center for Environmental Information, Inc.
Carol Werner, Director, Energy Program, Environment and Energy Study Institute
James C. White, Center for the Environment, Cornell University
Richard Wilson, Director, Northeastern Regional Center for Global Environmental Change, Harvard University

CONFERENCE COSPONSORS

Contributing Cosponsors

American Gas Association
E.I. Dupont de Nemours and Co., Inc.
Edison Electric Institute
Electric Power Research Institute
Global Climate Coalition
Harvard University, National Institute for Global Environmental Change, Northeast Regional Center
Motor Vehicle Manufacturers Association
National Hydropower Association
Ontario Ministry of Energy
Ontario Ministry of the Environment
Solar Energy Industries Association
United Nations Environment Programme
U.S. Department of Agriculture, Global Change Program Office
U.S. Department of Energy
U.S. Environmental Protection Agency
World Resources Institute
World Wildlife Fund - The Conservation Foundation

Cooperating Cosponsors

Agency for International Development, Bureau for External Affairs
Air and Waste Management Association
Alliance to Save Energy
American Association for the Advancement of Science
American Chemical Society
American Council for an Energy Efficient Economy
American Wind Energy Assoiation
Climate Institute
Cornell University, Center for the Environment
Environmental and Energy Study Institute
Environmental Defense Fund
Friends of the Earth
International Institute for Applied Systems Analysis
Izaac Walton League of America
Ministere de l'Environnement du Quebec
National Wildlife Federation, Environmental Quality Division
Natural Resources Defense Council
Princeton University, Center for Energy & Environmental Studies
Union of Concerned Scientists
U.S. Council for Energy Awareness
U.S. Department of Agriculture, Economic Research Service
University of Maryland at College Park, Center for Global Change
World Meteorological Organization

CONTENTS

THE POTENTIAL FOR NON-FOSSIL FUEL ENERGY SOURCES

THE POTENTIAL FOR REDUCING CARBON EMISSIONS THROUGH EFFICIENCY

THE POTENTIAL FOR NON-FOSSIL FUEL ENERGY SOURCES
INTRODUCTION

Richard Ottinger

Pace University Law School

The world is about to let the sunshine in and we feel the winds at our back as we approach the energy problems of the world today in the next decade. The advent of a new administration that is committed to renewable energy and efficiency, the passage of the Clean Air Act and the Energy Policy Act in the last session of the 1992 Congress indicate that the United States is finally going to resume its leadership role in advancing renewable technologies and efficiency in this country and around the world.

That is very heartening. There isn't a moment to spare. Despite the conflicts that exist within the scientific community about when and how severe global warming consequences are going to be, from the International Panel on Climate Change there appeared a worldwide consensus that climate change is indeed a real problem that we face and the consequences could be disastrous for all the countries of the world. The only prudent course for us to take is a course which protects us from those consequences in a no-regrets policy.

The question of how renewables are going to achieve their proper place in the energy perspective depends to a large extent on whether we can get the prices right and whether we can remove some of the barriers which exist in present law, which prevents renewables from being properly evaluated by all sectors of our industry, particularly the electricity sector.

It was very heartening and I think a lot of people don't know that the Rio declaration declares that the environmental costs of energy production are to be considered by the nations of the world and the externalities involved with the environmental damages are to be taken into account. This would be a major boost, if adopted, in advancing the economics of renewables. Many of them are at a point where, for many applications, they are economic today, and, for technologies like photovoltaics, the price is declining rapidly enough so that there are many niche applications that utilities can use and the uses will become much wider.

They are being widely used in Third World countries. They are being used to fill the needs for electricity in rural communities, where the costs of building transmission and distribution lines exceeds the cost of using photovoltaics.

The new Energy Act contains new tax incentives for production of electricity by renewables. It contains new incentives for utilities to engage in integrated planning, so that you compare different supply sources against demand systematically. All of that is kicking in at a time when we have a new administration with Clinton and Gore, both of whom understand these technologies and look upon them favorably.

Global Enery Strategies: Living with Restricted Greenhouse Gas Emissions, Edited by J.C. White, Plenum Press, New York, 1994

At the same time, we have an increasing number of poor nations throughout the world that are desperately seeking to feed their populations and for which the energy costs of development under traditional schemes are simply prohibitive. Many of these countries have no electricity distribution systems at all and for them renewable energy resources for agriculture, for electrifying and bringing light and refrigeration into their communities are most desirable.

Within the World Bank, for the first time, there is a real emphasis on environmental sustainability as a guiding principle for financing. The very aggressive and able head of the Environmental Division is asserting himself to see to that.

We are not done by a long shot, but the momentum, is going in the right direction. I think that, while we have our work cut out for us, there is much to encourage the efforts of those who believe that the future belongs to renewable energy resources and energy efficiency measures. We play our small part at Pace University, where we have an energy project, which has been advancing conservation and renewables for electric utilities in New York. It started about a year ago in Florida and most recently is starting in Michigan.

There are similar groups now formed throughout the country, spearheaded by not only ourselves but the Conservation Law Foundation in New England and the National Resources Defense Council. They are helping groups like the Southern Environmental Legal Center, a group in the Mid-Atlantic States; a very active Western group in the Northwest, a very active group in the Rocky Mountain area, and the Land and Water Conservation Group of the Rockies.

There is a citizen push to make conservation and renewables happen. It is working at the state and local levels and also pushing the national government to provide more sympathetic legislation.

This presents a very good climate for this Conference.

SOLAR POWER

Ann Polansky

Solar Energy Industries Association

The Solar Energy Industries Association has over 300 members which manufacture, distribute and install a wide variety of solar energy technologies: photovoltaics, solar water heating, space heating and cooling systems, and solar thermal electric and industrial systems.

Solar technologies have advanced to the point where they are ready for commercialization but they face significant market barriers. While there is strong evidence of widespread public support for solar energy, too often there is the perception that these technologies are either too expensive, not reliable, or that they are not technologically developed. Part of the reason I am here today is to dispel such notions. A wide variety of solar technologies are available today and have a multitude of applications to provide a wide range of energy services.

I will start with the basic high-temperature solar thermal technologies, which can deliver both heat and electricity. Examples are the parabolic trough, the dish-engine, and the central receiver.

The parabolic trough, for example, can heat water for use in industrial processes, or can create steam used to drive a conventional turbine. Figure 1 shows parabolic troughs at the facility in southern California formerly known as Luz International and currently operating as Kramer Junction Company. This facility, the only one of its type in the world, supplies about 350 megawatts of electricity to the grid, purchased by Southern California Edison-enough power for about 400,000 homes in southern California. A tube mounted in the center of each trough carries a fluid with very high heat capacity. Heat is transferred via a heat exchanger which then converts water to steam and turns a turbine. As an aside, I wanted to point out a that as a result of the volcano at Mt. Pinatubo last year, particulate matter in the atmosphere reduced the amount of available solar radiation, resulting in a 20 percent reduction in solar power production. As a result the facility had to apply for a special exemption from the government to allow it to rely more heavily on natural gas to produce power.

Figure 2 is a central receiver in Southern California, first constructed in a project known as Solar One. A consortium comprised of utilities, the solar energy industry, and the Energy Department is now in the process of upgrading and retrofitting this facility to incorporate new technological advancements such as a molten salt storage system. The mirrors that you see surrounding the tower are called heliostats, which reflect and focus heat from the sun onto the receiver tower where heat is converted to electricity. Several utilities have committed to

Global Enery Strategies: Living with Restricted Greenhouse Gas Emissions, Edited by J.C. White, Plenum Press, New York, 1994

FIGURE 1. THE PARABOLIC TROUGH

FIGURE 2. A CENTRAL SOLAR RECEIVER

cost-share the development and testing of this technology because they believe it could become a viable electricity generating technology in the next few decades.

In a solar dish-engine system, (Figure 3) in the center of the parabolic dish is a Stirling engine, an external combustion engine. DOE and the solar industry are now concluding a highly successful joint ventures program to demonstrate seven-kilowatt dish-engines all over the U.S. One system is installed and operating at the Pennsylvania Energy Office, which proves that solar energy is a viable technology on the Eastern seaboard as much as it is in the southwestern states. A follow-on joint venture, to demonstrate a larger, 25-kilowatt, utility grade dish-engine system is being initiated this year. Electric utilities are also very interested in the dish-engine technology and have expressed a strong interest in cost-sharing the 25-kilowatt joint venture.

FIGURE 3. A SOLAR DISH ENGINE

There are many uses for photovoltaics other than powering wristwatches and calculators. They have many uses in consumer products, but are also being investigated for their use in grid-connected applications. They are especially useful in remote locations for power supplies, communications, cathodic protections, battery charging, warning signs, water pumping and even restroom facilities.

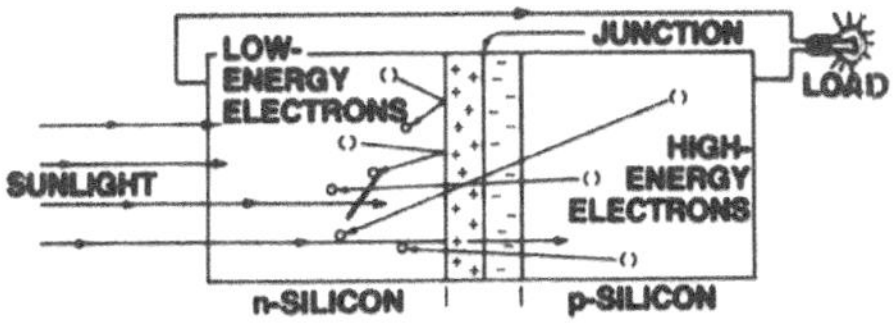

FIGURE 4. THE PHOTOVOLTAIC CELL

Some of you may not be familiar with the basics of photovoltaic technology. Figure 4 is a scematic showing how a PV cell works: photons from the sun excite electrons within PV cells, which then causes electron flow and electrical power is produced.

The home illustrated in Figure 5 is powered by photovoltaics. There are instances in which it is less expensive to install a PV system to provide residential power than to extend power lines from the grid. Pacific Gas and Electric's current policy is to install photovoltaic system for any additional electricity needs located more than a quarter of a mile from the existing grid. This policy is not based solely on environmental considerations -- it was found to be the more cost-effective alternative.

Pictured in Figure 6 is a water pumping system typical of many PV systems that have been installed throughout the world. This system pumps 5.7 cubic meters of water daily from a 45 meter well, supplying drinking water to a native American community in California. Two billion people in the world are still without access to electricity. In many places, a PV water pumping system is the only means available for obtaining groundwater. PV can also be combined with wind systems and with diesel power to form hybrid systems.

FIGURE 5. A PHOTOVOLTAIC POWERED HOME

I would like to move to an entirely different set of solar technologies: solar water heating systems. Most residential solar water heating system feature a flat plate solar collector. The solar water heating industry has been in a recovery period for past several years as a result of the discontinuation of residential tax credits which resulted in many solar water heating companies going out of business. Now we have an industry which is maintaining consistent market growth in the absence of a residential tax credit, an arguably healthier market condition. Solar water heating is especially competitive where electricity is being used to heat water.

Solar systems are used for pool heating. The United States consumes an enormous amount of energy to heat swimming pools -- several quads. Solar pool heating systems can drastically decrease that energy use while simultaneously offsetting carbon dioxide emissions and other pollutants. Fuel savings can result in a payback period of just a few years.

The high initial capital cost of solar technologies compared to the more conventional alternatives is a strong market barrier to solar commercialization. However, costs could be

FIGURE 6. A PHOTOVOLTAIC WATER PUMPING SYSTEM

reduced markedly by scaling up manufacturing, when there is sufficient market demand (Figure 7). SEIA, therefore, spends much of its energies in developing what we call "aggregated markets" for solar technologies. In increasing the demand, solar manufacturers can achieve the economies of scale needed to mass-produce and thus cut production costs. Solar energy costs can be significantly reduced through the proper implementation of a series of engineering and social choices. While much work has been done to develop the technologies themselves, less has been done to study how society values renewables and how those values can or should be integrated into standard decision-making processes. To forward these ideas, SEIA has employed nationally recognized economists to develop new, more appropriate approaches to the outdated methods utilities use to choose options for new capacity. We are also analyzing the effect that "green pricing," externalities, and carbon taxes would have on the cost-competitiveness of solar energy.

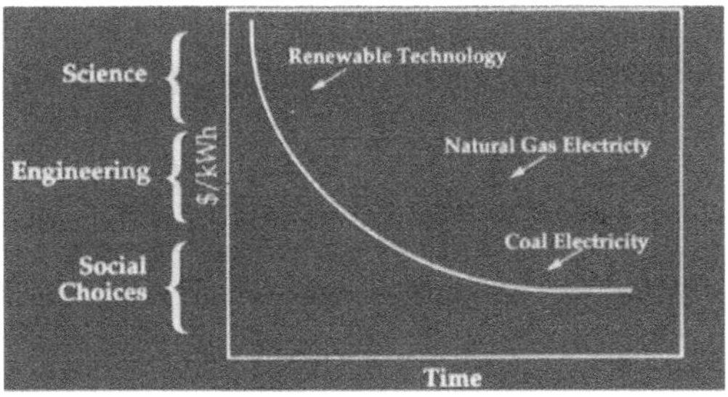

FIGURE 7. COSTS OF SOLAR TECHNOLOGY

One of the most prominent market barriers our industry faces lies in inequitable energy tax policies, depicted in Figure 8. Current tax policy allows a fuel-consuming energy production facility to take advantage of fuel-expensing. The bulk of the cost of a solar system is in its up-front capital cost.

If you ignore all but the bottom line of numbers in Figure 9, it shows that a solar facility still pays 10 percent more in taxes than a comparable gas fired power plant. The solar industry does have one tax credit, now permanently in place as part of the Energy Policy Act, which helps to ameliorate this situation. The business energy tax credit allows a commercial or industrial entity who purchases and installs a solar energy system to take a 10 percent tax credit. The Energy Policy Act also included a new tax credit -- a production tax credit for wind and some uses of biomass -- as well as a production incentive for utilities using renewable energy to produce power. These tax incentives are helpful but they do not go far enough. Major revisions in our tax code are needed before we have a business environment that allows all energy technologies to compete fairly for market share.

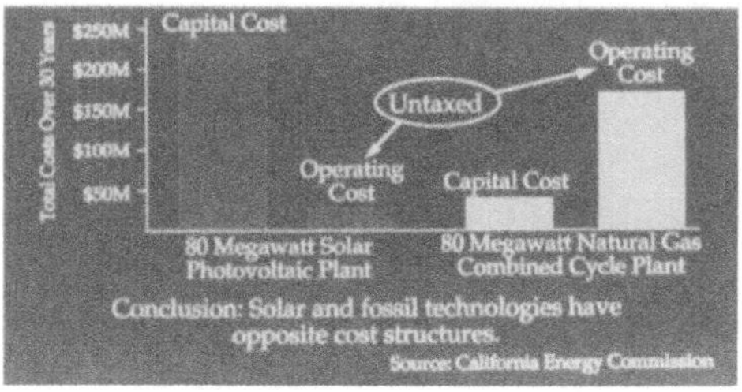

FIGURE 8. COMPARISON OF CAPITAL AND OPERATING COSTS SOLAR VS FOSSIL FUEL PLANTS

	Luz Solar Plant	Natural Gas Combined Cycle Plant
Federal Solar Tax Credits	$20,000,000	
State Solar Tax Credit	$20,000,000	
Sales Tax (6.5%)	$ 8,900,000	$ 1,800,000
Property Tax (1% year, 30 year life)	$60,000,000	$11,600,000
Net 30 year tax costs:	$28,900,000	$13,400,000

Conclusion: The combined Federal and State solar tax credits (both assumed to be 10%) significantly diminish the tax inequities, but the solar plants still pay over 10% more in taxes than the gas-fired plant of comparable size.

FIGURE 9. STATE AND LOCAL TAX PAYMENTS AND CREDITS OF A COMPARABLE SOLAR AND FOSSIL FUEL PLANT

The policies of the Reagan Administration are clearly shown in Figure 10, the federal R&D expenditures for renewable energy over the last 20 years. The solar R&D budget at DOE plummeted in the early 1980's, and in many ways the programs at the national labs are still recovering from these budget cuts. Much of the credit for keeping any semblance of solar R&D alive can be attributed to Scott Sklar (SEIA's Executive Director) for ensuring a minimum level of activity and for guiding the R&D programs to be relevant and useful to the solar energy industry. Recently, SEIA joined efforts with state organizations to develop a "Sustainable Energy Blueprint" which lays out a set of funding priorities for renewable energy and efficiency R&D programs as well as policies for reducing market barriers to sustainable energy.

My concluding message is simple. Renewable energy protects the environment while creating jobs. Two recent reports substantiate these claims. First, the Environmental Protection Agency released a report titled "Renewable Energy Electric Generation" which addresses the air pollution prevention potential of renewable energy technologies, and is, to my knowledge, the most quantitative, comprehensive analysis to date in this area. Copies are available from EPA's Office of Air and Radiation, Global Change Division. The second is a study conducted jointly by the Solar Energy Industries Association, the Alliance to Save Energy, and the American Gas Association, titled "An Alternative Energy Future."

Economic/energy models were used to predict environmental and employment benefits of an energy economy which relies heavily on a combination of natural gas, efficiency, and renewable energy. It predicts that in such a "sustainable energy future" between 200,000 and 400,000 new jobs would be created by the year 2010. This represents a net gain in jobs,

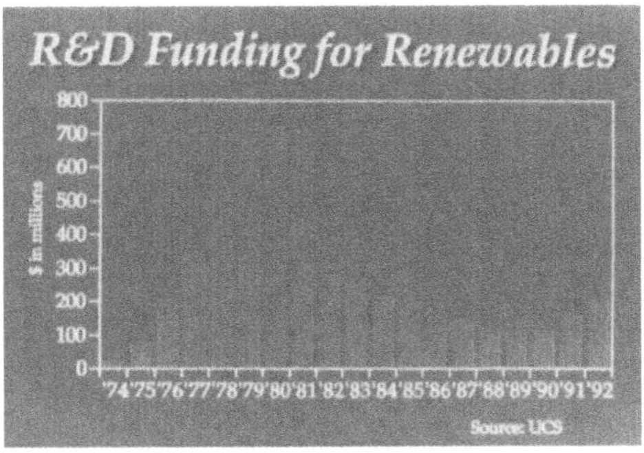

FIGURE 10: R&D FUNDING FOR RENEWABLE ENERGY

even accounting for jobs lost in the coal and other fossil fuel industries. This message is consistent with that of our Vice-President Gore: environmental protection offers economic opportunities. Environmental products and services represent a $300 billion market worldwide, and has enormous potential for growth. The message is simple. The U.S., through its national laboratories, is the leader in technological development and innovation. We risk losing our leadership role in the international marketplace by failing to take these environmentally friendly, innovative technologies to commercialization.

WIND TECHNOLOGY AFTER A DECADE OF DEVELOPMENT

Randall Swisher, Executive Director

American Wind Energy Association

INTRODUCTION

After a decade of development in California, wind energy technology has matured to the point where it is ready for widespread application to meet utility needs in many areas of the United States and the world. The inclusion of a 1.5 cent per kWh production incentive for wind-generated electricity in recently enacted national energy legislation could be the stimulus to extensive wind development.

Over the last decade, California windfarms have played a key role in the evolution of wind technology, with dramatic gains in both cost-effectiveness and technical performance since the first windfarms were established in 1981.

This progress has begun to be recognized by the U.S. utility industry and by public policy makers, and after a number of years in which major market activity for the wind industry was focused in Europe one can begin to see the opening of opportunities for wind technology in the U.S. in regions outside of California. Specifically, wind energy development is beginning to take place in the Pacific Northwest, the upper Midwest and the Northeast. The speed and scale of that development is still unclear, but this expansion of the market for windpower will be significantly influenced by federal and state policy decisions that will be made over the next few years.

CALIFORNIA WINDFARMS: A DECADE OF PROGRESS

Harnessing the wind for utility bulk power production proved to be much more challenging than renewable energy advocates expected, but windfarm performance has improved steadily since the first wind power plants began operation in California in 1981. In fact, the evolution of wind technology has been so rapid that there is now a significant gap between the perception of wind technology by many utility personnel and its current technical performance. That gap has been narrowed over the last few years due to the active efforts of knowledgeable utilities such as Pacific Gas & Electric and the Electric Power Research Institute.

Global Enery Strategies: Living with Restricted Greenhouse Gas Emissions, Edited by J.C. White, Plenum Press, New York, 1994

Production of wind-generated electricity in California has risen from 6000 kWh in 1981 to almost 2.8 billion kWh in 1992. Capacity factors have increased from an average 3 percent in 1982 to 20 percent in 1991, with turbines installed since 1985 averaging 26 percent. Some prime sites have exceeded 40 percent.

Availability factors have also steadily increased from a dismal 50-60 percent at the beginning of the decade to over 95 percent since 1985. Availabilities exceeding 98 percent are not uncommon at a number of wind power plants containing hundreds, even thousands, of wind turbines.

Despite lack of federal leadership throughout most of the decade, installed wind capacity in California reached 1, 619 MW by the end of 1991. Installations have been uneven, peaking to meet the expiration of the federal investment tax credit in 1985, then gradually trending downward as financeable Standard Offer 4 (SO4) contracts were completed in the following six years.

California's base of approximately 16,000 wind turbines, concentrated primarily in mountain passes at Altamont, Tehachapi and San Gorgonio, has provided a laboratory for technology development. The costs of wind-generated electricity have declined approximately 75 percent since the early 80's, primarily because of operational learning by windfarm developers and their O&M personnel. Costs of O&M have declined from approximately four cents per kWh to around a penny per kWh, although costs vary considerably depending primarily upon the age of the project. Older projects involving smaller turbines are generally more expensive to maintain than post 1985 projects. Capital costs have likewise declined from $2,500-3,000 per kW of capacity to around $1,000 (with considerable variation depending upon project specifics). Costs of existing projects are typically six to nine cents per kWh depending upon wind regime, size of the project, project financing and transmission investment required. Projects for construction in the mid-90's should be available for approximately $800 per kW of installed capacity with resulting costs of power in the 5-6 cents per kWh range for good wind regimes.

Wind development flourished in California primarily because of progressive state policy epitomized by the SO4 contracts. These contracts were designed by California regulators to be compatible with the characteristics and unique financing needs of new renewable technologies. Unfortunately, no other state stepped forward with such innovative policies, and wind development has been virtually nonexistent in any other state except Hawaii, which has only about 35 MW of installed capacity. In 1993, however, the wind industry has seen evidence that development will soon be occurring in a number of regions beyond California.

EMERGING MARKETS FOR WIND TECHNOLOGY IN THE U.S.

As the cost of wind-generated electricity continues to decline and utilities learn more about wind's improved technical performance, substantial progress toward wind energy development has been shown in three regions of the U.S.: the Pacific Northwest, the Upper Midwest and the Northeast. All three regions share at least a modest need for new sources of electric power, but the more important factors contributing to the movement toward wind are 1) utilities and state regulatory commissions that are "on the cutting edge" in their openness to new technology and 2) increased concern about the environment within each of the three regions, which has lead to more progressive energy policy decision making.

THE PACIFIC NORTHWEST

In the Pacific Northwest, the Northwest Power Planning Council (NPPC) is made up of two representatives appointed by each of the governors of the four states in the region (Washington, Oregon, Idaho and Montana), and is authorized by federal law to oversee power planning for the region. One provision of the law gives a higher priority to renewable energy development than to conventional power resources. The NPPC laid the foundation for wind development with a "confirmation agenda" designed to identify the steps necessary to foster cost-effective development of wind and other renewable resources in the region. Through a collaborative process involving utilities, regulators and renewable industry experts, the confirmation agenda resulted in recommendations for specific steps in the areas of resource confirmation, permitting, siting, and systems integration, and for research focussed on issues of concern to regional utilities such as cold weather operations and icing.

Although the NPPC has limited direct authority over the region's utilities, the regional planning process helped develop a consensus on the Northwest's energy future, including a major emphasis on renewable energy development. Many utility personnel retained a negative perception of wind technology based upon project failures such as the federal MOD-2 turbines that were erected in the Northwest, but those perceptions began to fade as utility managers learned more about the performance of newer generations of commercial wind turbines.

Last year witnessed dramatic progress toward wind development in the region as four of the largest utilities in the Northwest (Puget Sound Power & Light, PacifiCorp, Portland General Electric and Idaho Power) announced a commitment to a 50 MW project at Rattlesnake Mountain in eastern Washington in cooperation with U.S. Windpower. Project construction is expected to begin in 1994 and be completed by 1996. In addition, in June, 1992, the Bonneville Power Administration (BPA), a federal power marketing administration that serves about half the load in the region, announced a solicitation for up to 50 MW of windpower, with an expected on-line date of mid-1995. Bonneville received ten proposals totalling over 270 MW of wind capacity, selected two of them for the "short list" for negotiations for power purchase, and is negotiating with a number of others to provide transmission and other services to the utility partners involved.

With the announced 1996 closure of the Trojan nuclear plant and diminished hydro-based generation because of operational changes required to protect the region's remaining salmon, the Northwest is facing a rapidly growing need for power. As a result, the commitment to wind is expected to grow over the next few years. By way of example, PacifiCorp has announced plans to purchase up to 500 MW of wind capacity by 2001 (depending upon cost-effectiveness), and Portland General Electric, having seen the effectiveness of the BPA Wind Solicitation, has completed a draft renewable energy solicitation that will be used to purchase primarily wind or geothermal capacity. Publicly owned utilities in Washington state have also established the Conservation and Renewable energy System (CARES) as a joint operating agency which will finance, design and implement conservation and renewable energy projects for its members. Wind projects are high on its list of renewable options, and CARES was one of the utilities to submit a wind proposal to BPA.

AWEA held Windpower '92 in Seattle, Washington, between October 19-23, for the purpose of showcasing wind technology for decision makers in the region, adding to the momentum that is already evident.

The Bonneville Power Administration's Wind Solicitation has brought about a fundamental change in the way wind is perceived by utilities in the Pacific Northwest. Bonneville's offer to purchase up to 50 MW of wind capacity from teams of utilities and developers forced utilities in the region to take a close look at the current status of wind technology. Prior to the solicitation, most utilities in the region considered wind an R&D technology most appropriately treated through very small pilot projects. Virtually every major utility in the region is now involved in actively considering wind projects to provide commercial power.

THE UPPER MIDWEST

In the last year, commercial progress by wind technology in the Upper Midwest has been almost as dramatic as in the Pacific Northwest. Minnesota, which has considerably more wind potential then California, has led the way in the region through years of resource assessment and other policies intended to make the state the center of Midwest wind energy development. Leaders in the state legislature have sponsored a series of pro-wind measures, including exempting wind projects from the state property tax and incorporating the environmental benefits of wind projects into the price paid to project developers. Although the environmental benefits of wind development have been an important justification for the legislation, the key appears to be economic development: making Minnesota the heart of the emerging wind industry in the region, and helping to create new and enduring jobs for the 21st Century.

Minnesota wind development received a boost on October 1, 1991, when Northern States Power (NSP), the state's largest utility, announced plans to construct a utility-owned 10 MW windfarm by 1996. The state government and the environmental community responded by pressuring the utility for a larger commitment of at least 50 MW. In August NSP responded by announcing a commitment to at least 100 MW of wind capacity, with 5 MW in 1993, 20 MW more in 1994, and at least 75 MW more by the end of the decade.

A much smaller project, but significant nevertheless, is the small five-turbine windfarm that began operation in April serving the Marshall, Minnesota, municipal utility. The contract with developer Minnesota Windpower provides the utility with only one million kWh annually, but small municipal or cooperative utilities could represent a growing market for wind. In total, municipal and cooperative utilities serve about 25 percent of the U.S. demand.

Minnesota is not the only state in the region that has shown interest in wind. The Wisconsin Public Service Commission modified the long range plans of the state's utilities to require 811 MW of renewable energy by 2010, with wind being one of the most likely renewable options.

One of the most significant steps in the Midwest was taken when U.S. Windpower formed a joint venture with a subsidiary of Iowa-Illinois Gas & Electric to develop wind projects within a multi-state region. The joint venture, WindRiver Power Company, is now actively pursuing projects throughout the Midwest.

THE NORTHEAST

Firm commitments to wind projects in the Northeast lag behind the other two regions, but there has been clear progress in the last year, nonetheless. New England Power (NEP) has been in the forefront of many changes sweeping the utility industry, including interest in renewable technologies. In December, 1991, NEP issued a "Green Solicitation" for 45 MW of renewable capacity, specifically identifying wind as one of the priorities for the solicitation.

In response to the solicitation, Green Mountain Power (GMP) proposed three specific projects in Southern Vermont and Western Massachusetts totalling 40 MW. GMP has been the most active utility in the Northeast in terms of involvement with windpower, having monitored wind resources for a decade, and invested in California windfarms through a subsidiary.

According to reports published by the Battelle Pacific Northwest Laboratory for the U.S. Department of Energy, Maine has the largest wind resource in New England. Currently, Endless Energy, a small developer, has a research contract with Central Maine Power (CMP) for 15 MW that must be developed by 1994. U.S. Windpower has also been active in the state, proposing a 50 MW project in western Maine to CMP.

The state of New York has also begun to take steps to harvest more of its large renewable energy potential. Earlier in the year, the state updated its energy plan, incorporating a proposal that requires the state's utilities to purchase a total of 300 MW of renewable energy capacity by 1994, with projects on line by 1998. The proposal has been referred to state regulators for implementation.

AWEA'S POLICY AGENDA TO CONTINUE WIND'S U.S. MARKET EXPANSION

AWEA advocated four part program in 1992 designed to open new markets for wind energy technology:

I. Production Incentives

AWEA's priority has been tax equity--specifically, providing incentives for wind energy development that create more of a "level playing field" with conventional technologies, which receive billions of dollars in tax subsidies annually. The vehicle chosen was a production incentive for each kWh of wind-generated electricity for new projects.

Although few observers gave the production incentives campaign much chance of being successful when it was initiated, a 1.5 cent per kWh production incentive for wind became law through the National Energy Policy Act of 1992. The production incentive will go into effect at the beginning of 1994 and should provide a significant impetus to the market.

II. Transmission Access

The bulk of the U.S. wind resource is in the Great Plains, far removed from load centers on either coast. Access to transmission lines to bring wind-generated electricity to load centers is essential if we hope to see wind energy reach anything close to its potential.

The National Energy Policy Act included transmission access provisions that were actively supported by AWEA that would substantially increase access to transmission and would represent a boost to the wind energy industry.

III. Enhanced Research and Development

The federal wind program has never fully recovered from the 90 percent reduction in funding that it suffered during the 80's. Despite a doubling of the wind budget in the last two years, the federal commitment to wind is only $24 million for FY 93, about half the level of the wind technology development efforts that have been funded in western Europe. Sophisticated R&D will be essential for wind technology to achieve the cost reduction targets that are feasible.

IV. Regulatory Reform

Wind and other intermittent renewable technologies are fundamentally different than conventional technologies. The decision making criteria used by most utilities and utility regulators reflect the characteristics of conventional technologies and don't take into consideration either the full costs of conventional technologies (externalities) or all the benefits that renewable technologies bring to utility systems. Unless there is a systematic redesign of the utility resource planning and selection process, wind and other renewables will continue to be disadvantaged in the energy market.

Regulatory reform has focused primarily on energy efficiency and demand-side management in the last five years, but with so much of that agenda having been accomplished, more attention is being turned to the impact the regulatory process and utility resource planning have on renewables.

With progress on tax equity and transmission access, regulatory reform will become the heart of AWEA's political agenda.

CONCLUSION

Since the first windfarms were developed in 1981, wind technology has made tremendous progress in terms of both technical performance and cost-effectiveness. Despite this progress, the market for wind in the U.S. has been confined almost entirely to California. However, 1992 has provided clear evidence that the market for wind technology is emerging in at least three other regions of the U.S. This trend could be markedly accelerated by more effective policy leadership at the federal and state levels, including an active partnership with the wind industry to win widespread commercial development of the most cost-effective of the solar electric technologies.

THE POTENTIAL FOR BIOMASS ENERGY AND THE IMPLICATIONS FOR CLIMATE CHANGE

David C. Rinebolt

Legislative Director
Office of Representative Collin C. Peterson

INTRODUCTION

Using wood and other types of biomass for fuel can slow the buildup of carbon dioxide and other 'greenhouse gases' in the atmosphere. Growing biomass removes carbon dioxide from the atmosphere and converts it to carbon stored in the plant fiber. As long as wood used for fuel is replaced by new wood growth a carbon cycle is created and there is no net increase in the amount of carbon dioxide released. To the extent that wood replaced by new growth is substituted for fossil fuels there is a net reduction in the amount of carbon dioxide emitted into the atmosphere.

Using biomass as a substitute for fossil fuels is one way of mitigating the prospective increases in concentrations of carbon dioxide and methane in the atmosphere and the associated climate changes. It also offers a way to utilize the large volumes of biomass now available and the much larger volumes that would become available if we can increase forest areas, increase timber growth on existing timberland, grow short rotation woody crops and herbaceous energy crops, and plant trees in urban areas and shelterbelts.

THE EVOLUTION OF BIOMASS ENERGY IN THE UNITED STATES

The use of wood for fuel has a history as long as mankind. Archeological evidence indicates the use of wood fuel in man's earliest settlements. Wood was the dominant fuel in this country into the period of the industrial revolution, and remains the dominant fuel in many of the world's less developed countries.

Wood use for fuel in the United States probably peaked in the 1870's. It was the primary fuel used by households for cooking and heating and was also an important source of energy for manufacturing and transportation industries. After this consumption peak, wood was gradually displaced by coal, which combined a higher BTU value with lower production and

---

Reprinted by permission from American Forests. Vol. 2: In Press.

further diminished the importance of wood in the domestic energy mix. The downward trend transportation costs. The discovery and exploitation of petroleum and natural gas resources in the use of wood for fuel continued with some fluctuation until 1972 (Skog and High 1989).

Starting in 1973, in response to the price increases on crude oil imposed by the Organization of Petroleum Exporting Countries (OPEC), the trend reversed and a rapid increase in the use of wood for fuel began which lasted through the mid-1980's. The use of wood for energy continues to grow, albeit at a slower pace than several years ago.

Historically, most of the wood used for fuel has been roundwood, i.e. round sections of trees cut for use as fuel from forests, urban areas, and other sources. Since 1972 the use of roundwood for fuel has increased from about 500 million to over 3 billion cubic feet - a 600 percent increase. Much of this increase occurred in the residential sector, primarily for space heating. Limited but growing volumes of roundwood were used in manufacturing industries, utilities, and in commercial buildings.

Round fuelwood now accounts for about a fifth of the timber harvested from domestic forests. Most of this harvest, however, comes from sources such as rough, rotten, and salvable dead trees; tops and limbs; noncommercial species; trees under 5 inches in diameter; and trees from woodlands, fence rows, and urban areas that are not commonly utilized by the primary wood products manufacturing industries (USDA Forest Service, 1990).

In addition to roundwood, the equivalent of about 1.5 billion cubic feet of sawdust, slabs, edgings, chips, veneer clippings, and other solid wood byproducts of wood manufacturing plants is used as fuel, primarily in the source plants. These residues are also sometime processed into pellet fuels for wholesale and retail sale. Large volumes of bark, spent pulping liquor from wood pulp mills, discarded wood and wood fiber such as lumber from demolished buildings, and waste paper are also consumed as fuel (USDA Forest Service, 1990).

Most of the growth in the consumption of fuelwood since the early 1970's has taken place in households and in wood manufacturing plants. Significant growth has also taken place within the commercial, industrial, and utility sectors outside of the traditional forest products industry. The rapid increase in the use of wood for fuel in the last couple of decades was caused by a number of forces. These include the crude oil price increases of the 1970's imposed by the member countries of the Organization of Petroleum Exporting Countries (OPEC), environmental and energy legislation, improvements in wood burning technology, and the availability of large volumes of wood suitable for use as fuel.

The actions of OPEC also stimulated the resurgence and massive expansion of the domestic ethanol industry. The internal combustion engines of many early automobiles were actually designed for ethanol, which was more common than gasoline. Gasoline was viewed as an unwanted byproduct of kerosene production. Interaction between the fledgling automotive industry and the oil companies soon changed the situation.

In 1973, and again in 1978, OPEC imposed controls on the volume of crude oil that could be exported by the member producing countries. These controls constrained the supplies of crude oil available in world markets and made it possible to impose and sustain large increases in the price of crude oil. As demands shifted to other fossil fuels there were substantial and related increases in their prices. Between 1973 and 1981, the composite price index for fossil fuels in the United States rose from 80.4 to 292.0 measured in constant 1982 dollars. Most of this rise reflected crude oil prices which increased more than seven times. The price indexes of natural gas and coal measured in constant 1982 dollars increased from 40.6 to 191.0 and 73.7 to 125.4 respectively. In contrast stumpage (standing timber in forests) prices and log prices in this time period remained about the same when measured in constant dollars. Further, much of the additional wood used for fuel came from very low cost sources such as plant byproducts and trees not commonly used by primary wood manufacturing industries. Likewise, corn supplies were readily available at fairly low prices due to continuous overproduction.

The rise in fossil fuel prices relative to wood and other biomass feedstocks was surely a major force behind the large increase in the use of biomass for fuel that took place in the 1970's and early 1980's. Since the mid-1980's there have been fairly large declines in crude oil and other fossil fuel prices although the composite index still remains substantially above the levels of the early 1970's. Again, wood prices haven't changed much but there has been some decline in the use of round fuelwood. This drop has apparently been concentrated in households where fossil fuels have replaced some of the wood used for heating. This may in part reflect the much greater convenience of using fossil fuels.

Legislation was another force behind the rapid growth in the use of wood for fuel in the 1970's and early 1980's. There were three particularly important pieces of Federal legislation -- the Clean Water Act ("Federal Water Pollution Control Act Amendments of 1972"), the Clean Air Act Amendments of 1977, and the Public Utility Regulatory Policies Act of 1978 (PURPA). The Clean Water and Clean Air legislation was both implemented and supplemented by State and local government through laws and regulations.

The pollution control and abatement requirements of the Clean Water Amendments caused an increase in the use of spent pulping liquors, which contain large amounts of lignin, as a fuel in the wood pulp industry. The use of these 'black' liquors for fuel rose from 62 million tons in 1972 to 78 million tons in 1986 (American Paper Institute 1991).

Requirements for control of smoke and particulate emissions in the Clean Air Act affected the wood product industries that were burning wood wastes such as sawdust, slabs, edgings, bark, veneer trimmings, and shavings in teepee burners and other disposal devices. These requirements provided indirect incentives for using wood wastes in products such as pulp or as fuel. In the solid wood processing industries, for example, the use of wood and bark residues grew from 20.6 million tons in 1977 to 24 million in 1981 (Skog 1987).

State and local governments have been passing laws and promulgating controls on the disposal and recycling of wastes including wood and wood fibers (paper and paperboard). In some areas this has resulted in the increased use of waste wood and fiber for fuel. Although the volumes involved at this time are small the potential is large, both in terms of further legislation and controls and the amount of wood and fiber available. Wood waste has the potential to become an important source of fuel.

Until the late 1970's, the use of wood for energy was primarily limited to the wood products industries and in residential wood stoves and fireplaces. At that juncture, the second price increase imposed by the crude oil exporting countries triggered legislation to move American energy more toward 'alternative' fuels. The Public Utility Regulatory Policies Act of 1978 did this by providing a regulatory framework and a stimulus for the production of electrical energy from non-traditional energy sources by companies outside the utility industry. This legislation, and the market forces resulting from increases in fossil fuel prices, were effective in developing an independent power industry and in bringing wood into use in the utility sector.

The most recent data indicates that 5,435 megawatts of wood-fired capacity is now connected with the utility grid (Rader 1990). Only 250 megawatts of that capacity is utility owned; the remainder is generated by independent power producers or cogeneration at industrial facilities.

Federal legislation to stimulate the use of alternative fuels and market forces resulting from higher fossil prices have also brought about significant increases in the consumption of wood fuel in economic sectors outside the utility and wood processing industries. Further, and perhaps more important, this legislation and the market forces resulting from increased fossil fuel prices have provided the incentive for a technological renaissance in the use of wood for fuel.

FUTURE BIOMASS SUPPLIES FOR ENERGY USE

One of the major forces behind the rapid increases in the use of wood for fuel that took place in the 1970's and early 1980's was the ready availability of large volumes of low cost wood. This wood came from three basic sources -- the residues or byproducts of the wood manufacturing industries: roundwood cut from trees not commonly used by the primary wood manufacturing industries (nongrowing stock sources) such as rough, rotten, and salvable dead trees; tree tops and limbs; noncommercial tree species; trees under 5 inches in diameter; and trees from woodlands, fence rows, urban areas, and other similar sources: and roundwood cut from trees that are commonly used by the primary wood manufacturing industries i.e. growing stock on timberland as defined by the U.S. Forest Service. There are two additional sources of wood and wood fiber that while not used in large volumes at the present time have a big potential for use as fuel -- short rotation woody crops and wood and wood fiber from urban sources. Although all of these major sources have the potential to supply large additional quantities of wood or wood fiber for fuel the outlook for each is different.

There is about 307 million acres of cropland in the United States that have the combination of fertility, rainfall, and slope that make them suitable for the production of short rotation woody crops. There is an additional 85 million acres of pasture and forest land that have good to moderate potential for conversion to cropland. Of all the suitable lands some 225 million acres have the capability to grow 5 or more tons of dry woody biomass per acre per year with current technology.

Estimates prepared by the Soil Conservation Service and published in its recent Resources Conservation Act Appraisal indicates that the United States could idle 129 million acres of current cropland and still meet its needs for crops until 2030 (USDA Soil Conservation Service 1990). Future land availability for woody crop production would thus seem to be primarily a function of economic opportunity and technological development.

In the last couple of decades there has been a substantial amount of research, mostly carried on by the Oak Ridge National Laboratory, on the culture of short rotation woody crops and their conversion to fuel. Because the use of the research is still in the early stages there is not much of a basis for estimating the area of cropland that might be converted in the future and the resulting wood production. With present technology about 5 tons of net dry biomass could be produced per acre per year; with the improved technology that is attainable as much as 10 tons per acre could be produced.

This all suggests that there is a very large potential for producing fuel through short rotation intensive culture woody crops. The wood produced could be converted to liquid fuels such as ethanol or used to produce energy through direct combustion. It can also be used to manufacture wood pulp or composition products like particle board.

Some of the solid wastes originating in urban areas is used as fuel. For example, in 1986 these wastes were the source of fuel for about a tenth of the non-utility electric capacity based on biomass fuels (Rader 1989). One company in New York is using clean wood waste -- tree trunks and trimmings, wood pallets and other shipping materials, and wood construction and demolition debris as fuel for a 50 megawatt generating plant on Long Island (Rader 1989).

In total, however, only a small part of the wood and other convertible wastes from urban areas are being utilized. In 1988, for example, there was nearly a billion tons of urban wastes hauled away to landfills. This included 35 million tons of solid wood products, 395 million tons of paper and fiber, and 170 million tons of yard wastes (World Resources Institute 1992). Although part of this is not useable because of its location, small quantity, or composition it is obvious that wood and wood fiber urban wastes represent a very large source of biomass that could be used for energy. It is also a source that is likely to get bigger as population and economic activity increase in the future.

Utilizing the urban wood wastes for energy would have substantial associated environmental and economic benefits. It would reduce the release of methane, an important greenhouse gas, that results from the decomposition of wood and wood fiber in land fills. It would also reduce the costs of storing wastes and of the physical space needed for land fills.

CONCLUSION

It is evident that there are now large volumes of wood and wood fiber that are potentially useable for fuel. It is also evident that if the opportunities to expand forest areas, increase timber growth on timberland, grow short rotation woody crops, and plant trees in urban areas and shelterbelts much larger volumes will become available in the future.

4.5 billion cubic feet of wood plus large quantities of bark and spent pulping liquors, and some urban wood wastes are now being used as fuel. Because this wood and wood fiber is replaced by new wood growth (net annual growth is substantially above removals) there is no net addition to the amount of carbon dioxide going into the atmosphere. Further since nearly all of the wood is used in place of fossil fuels the amount of carbon dioxide going into the atmosphere is reduced by some 6 to 8 percent. Thus it seems clear that increasing the use of wood for fuel can have signicant impacts in reducing the prospective increases in concentrations of carbon dioxide and the associated climate changes. There would also be significant desirable effects on the society, the economy, and the environment.

Great progress was made in the use of wood for fuel in the 1970's and early 1980's. This resulted from increases in prices of fossil fuels relative to wood, legislation designed to protect the environment and encourage the use of renewable fuels, and research to improve the technology of using wood for fuel. In view of the large supplies of low cost wood and prospective shortages of some fossil fuels further price incentives seem inevitable. However, sustaining the other forces will require action.

There is surely room and need for further legislation to protect the environment. Among the promising possibilities would be legislation to further encourage the use of renewable energy sources and the utilization of wood wastes both in forests and in urban areas. The removal or relaxation of regulatory barriers that limit the use of wood for fuel could also contribute in important ways.

REFERENCES

American Paper Institute. 1991. U.S. Pulp and Paper Industry's Energy Use - Calendar Year 1990. New York, New York.

Energy Information Administration (EIA). 1989. Estimates of Biofuels Consumption in the United States During 1987. CNEAF/NAFD89-03. Office of Coal, Nuclear, Electric and Alternate Fuels, Energy Information Administration, Washington, D. C.

Meridian Corporation. 1989. Characterization of U. S. Energy Resources and Reserves Prepared for U. S. Department of Energy under Contract No. DE-ACO1-86CE30844. June 1989. Alexandria, VA. National Wood Energy Association (NWEA). 1988. State Biomass Statistical Directory. Washington D. C.

-------(NWEA) 1990. "Federal programs for research & development of biomass and municipal waste technology," in Biologue, Vol. 7, No. 1, February/March 1990. Washington D. C.

Rader, Nancy; K. Bossong, J. Becker, D. Borson, and C. Manuel. 1990. The Power of the States - A Fifty-State Survey of Renewable Energy. Public Citizen, Washington D. C.

Rader, Nancy; K. Boley, D. Borson, K. Bossong, and S. Saleska. 1989. Power Surge The Status and Near-Term Potential of Renewable Energy Technologies. Public Citizen, Washington, D.C.

Skog, K. and C. High. 1989. "Current and projected wood energy consumption in the United States," in Biologue, Vol 6, No. 2, April/May 1989. National Wood Energy Association, Washington D. C.

U. S. Department of Agriculture. 1980. A National Energy Program for Forestry. USDA Forest Service, Misc. Pub. No. 1394. Washington D. C.

U.S. Department of Agriculture, Forest Service. 1990. An Analysis of the Timber Situation in the United States: 1989-2040. Gen. Tech. Rep. RM-199. Ft. Collins, CO: Rocky Mountain Forest and Range Experiment Station.

U.S. Department of Agriculture, Forest Service. 1989. Forest Statistics of the United States, 1987. Res. Bull. PNW-RB-168. Portland, OR: Pacific Northwest Research Station.

U. S. Department of Agriculture, Soil Conservation Service. 1989. The 2d RCA Appraisal: Soil, Water, and Related Resources on Nonfederal Land in the United States: an Analysis of Conditions and Trends. Rep. 242-141/03004. Washington, DC: U.S. Department of Agriculture.

U. S. Department of Energy (DOE). 1988. Five Year Research Plan 1988-1992 Biofuels and Municipal Waste Technology Program. DOE/CH10093-25, DE880D1181. Biofuels and Municipal Waste Technology Division, Office of Renewable Energy Technologies, Washington D. C.

-------(DOE). 1990. The Potential of Renewable Energy - An Interlaboratory White Paper. Prepared for the Office of Policy, Planning and Analysis, under Contract No. DE-AC02-83CH10093. March, 1990. Solar Energy Research Institute, Golden, CO.

World Resources Institute. 1992. Environmental Almanac. Boston. Houghton - Mifflin.

Wright, L. L., J. H. Cushman and P. A. Layton. 1989. "Dedicated energy crops - expanding the market by improving the resource," in Biologue, Vol 6, No. 3, June/July/August 1989. National Wood Energy Association, Washington D. C.

HYDROPOWER

Linda Ciocci

National Hydropower Association

This forum comes at a very, very opportune time. As others have mentioned, these types of forums help to provide the type of leadership needed to move the national agenda forward. Each and every one of us in this room has a real role to play in expanding the use of renewable energy. And, of course, hydropower is very much a part of that national agenda.

We are coming to Washington at a very exciting time, as we see a new administration getting ready to take over. As many of you may know, Mr. Clinton has already announced his energy strategy for the future which includes increasing energy efficiency, conservation, and expanding renewable energy sources – including hydropower. The effort to triple, by the year 2010, the renewable energy production in this country, is proposed in a blueprint that has been provided to Mr. Clinton from a group of renewable environmental organizations.

Hydropower is not mentioned as part of that effort because it is considered a commercially mature renewable energy source. Many people, unfortunately, assume that hydropower is something that will always be here but my message to you today is that all of us need to work together to preserve the hydropower that we currently have in this country because it is an environmentally benign technology.

There is some concern that hydropower is the ugly stepchild. Some environmental groups have not embraced hydropower and specifically see some real problems with it in the future. They have worked very strenuously in Congress to try to stop hydropower development but we are working very hard to try to keep hydropower a very important part of our energy's future.

Hydropower is commercially economically viable. It is a ready renewable. It is something that we can put into place today and it is available in almost every state in this country and in 86 countries in the world.

The technology is simple. Water, from a height, is forced through a turbine that drives a generator and produces power that immediately moves to the power grid. It is economic; it costs on the average of 1,800 to 2,000 dollars per kilowatt hour to develop new hydropower and a hundred dollars per kilowatt hour to upgrade existing hydropower capacity. Hydropower ranks third in electricity generation in the United States. It makes up most of the renewable energy production today; 88,000 megawatts. This includes 17 gigawatts of pump storage, 64 gigawatts of conventional hydropower and 7 gigawatts of small scale hydropower, which we define as 30 megawatts and below.

Global Enery Strategies: Living with Restricted Greenhouse Gas Emissions, Edited by J.C. White, Plenum Press, New York, 1994

It produces 14.5 percent of the total world capacity, as it is developed in 86 countries. It is notable that 50 percent of the hydropower we currently have in the country is from private projects, all of which will require relicensing in the next 20 years.

There are many benefits from hydropower. In terms of this conference and the topic of discussion, global warming, there is tremendous benefit in expanding hydropower to deal with global warming concerns because it has no air emissions and it has no hazardous waste products.

It is also multipurpose in the fact that it provides both flood control, water supply and irrigation. It provides recreational opportunities for fishing, boating, hiking, white water rafting and camping. There are tremendous examples around the United States where the hydropower industries have provided recreational areas, many places where these opportunities would not be available without the project development.

It also has a tremendous economic return, both to communities and the nation overall. For hydropower, because it is a renewable water resource that we are using, there are no fuel imports. It helps to lower our trade imbalance, and reduces foreign control over the economy. There is price stability for the consumer in that it is inexpensive.

There are high up-front costs for development, but once you have developed it, there are no fuel costs and there are low operation and maintenance costs. Because many of the hydropower projects we have in this country are already developed and paid for, they help utilities keep down their overall operating costs.

This keeps cost down for the consumer, and also it provides recreational facilities that create job opportunities in local communities, producing a direct revenue return. A great operating benefit is the ability for these generators to immediately go on line, which helps in problems of peaking and voltage control.

Unlike some of the other renewable technologies, where the impediments for further expansion are more technical or economic, the real problems with hydropower rest primarily in societal concerns because of the environmental mitigation issue.

There is a perceived environmental harm in the use of this technology. But we have to remember that there is a natural effect on the ecosystem when you develop anything, hydropower, a road, a bridge, or a housing project, - all impact the habitat.

Hydropower does, indeed, impact the environment. There is no question and we don't deny that. But we have discovered through new technologies and a new understanding of this technology, that you can mitigate through proper planning and siting and good research and technology, the effects that the hydropower project may have.

Probably the biggest concern, raised during the energy debate and also to be raised in the endangered species debate, is the fish question. There is an impact to fish habitat and the industry is working extremely hard to deal with it. We have spent millions and millions of dollars for upstream and downstream passage, on the development of fish ladders and appropriate spillways that enable fish to migrate.

More can be done in this area. We believe that it is not impossible to protect the environment, protect fish habitat, and protect wildlife habitat at the same time that we are continuing to develop hydropower.

One of our greatest concerns is the lack of consensus among the scientific community on how best to meet the problems we face. It is extremely difficult to work through the regulatory process, which is very complex. There are a number of agencies claiming jurisdiction over hydropower projects and there are differences between federal and state agencies and also between federal agencies themselves.

There are 41 different statutes that govern hydropower. The regulation process is absolutely mind-boggling. Preparing for this paper, we developed a chart of the regulatory process. This chart is three specific pages, legal size, readable only with a magnifying glass. We were hoping that we could take that three page, legal size document and somehow put it in a simple, understandable form where you would be able to see the various steps that are required from the preconsultation process to the actual licensing process. We were unable to do this.

Then there is the opposition from various interest groups that are impacted by the project. These groups are local, regional and federal and they rarely reach consensus. One of the things that we have tried to do in the industry is try to move toward a settlement concept, where we bring all these competing interests and federal and state agencies together in one room to work toward an agreement on a license. We are discovering that this process, even though it is time-consuming, can be very, very successful. The Congress and federal and state agencies are forcing utilities and operators to move into this settlement process, but some national environmental groups are working in opposition to that, intervening at the Federal Energy Regulatory Commission (FERC).

This lengthy licensing process takes anywhere from five to seven years. There are high up-front licensing and construction costs but the long term life of the project makes the technology economically viable.

The issue of the lengthy licensing process, the need to reach consensus, the inability of the scientific community to come together on the types of studies that need to be done, all pose an uncertainty with the project and questions as to whether the project is going to get a license. This also affects the project financing.

Many financing banks will not get involved in these projects because of the uncertainty and, again, that affects long term utility planning. Is it economically viable for them to put that much effort into conducting the studies if the license itself is uncertain?

Many of the groups that have opposed hydropower over the past few years are concerned that all that the hydropower industry would like to do is build more dams, which will significantly impact the nation's waterways. The truth is that there are 63,000 existing larger dams, 80,000 dams total in the United States and only 3 percent have hydropower facilities. Thirty-five percent of the dams were built for recreation purposes. They have nothing to do with hydropower generation; 18 percent are for stock and farm ponds, 15 percent are for flood control; 12 percent are for public water supply; 11 percent are for irrigation and only 2 percent are for hydropower.

We could increase hydropower's capacity by 22 gigawatts, just by building projects on existing dams. That is a tremendous amount of potential. Thirteen gigawatts could be developed in an acceptable cumulative impacts methodology – if new methodologies were developed. That means that in addition to the 22 percent, we could add 13 percent more if we spent more time and dollars in trying to find the best way to mitigate some of the environmental issues and if we could agree on procedures, etc.

These 25 to 40 billion kilowatts from operations and equipment enhancements would equate to 55 million barrels of oil.

We are greatly encouraged by the Clinton Administration, expecting that they will have an understanding of the issues, a willingness to get beyond the misconceptions, and move forward with a true agenda in the renewable area. This will require that we aggressively develop policies that promote renewables. We have an opportunity that we must capitalize on with a new administration that recognizes the importance of renewable technologies. We need to develop balanced policies that address the environmental concerns and the need for electricity.

We need to be very, very realistic about how we can move forward to develop hydropower in this country and protect the environment. These are not mutually exclusive goals.

We have to streamline the licensing process and consolidate the environmental review process. We have to make these settlements, that the FERC is moving forward with, work. The only way that we are going to be able to do that is to work together. We had developed a number of proposals in the National Energy Bill that would have streamlined the licensing process. We lost each and every one of them in the energy debate.

There was a tremendous proposal, sent up in the national energy strategy under the Bush Administration, that would have had an incredible benefit to the hydropower industry by consolidating the various review processes at the federal and state level.

We need to move forward with that agenda in order to see the full hydropower potential develop. Industry is not going to continue to develop hydropower if we end up having to protect and fight for just what we have got. We won't see any new development.

The last issue is that we need to increase the research and development efforts. A lot of money is being spent in the solar area. Hydropower has very little money being spent at the federal level right now. In fact, we have one utility, PG&E, that spends more money on R&D in the hydro area than the entire Federal Government because they believe in it and they recognize its full potential.

We need to get the Federal Government to focus on this more in terms of equipment, operations, maintenance and the biological study area, to improve environmental mitigation. There is much that can be done. We have talked to the new administration and I think they recognize the role that R&D can play in improving hydropower in the country and are willing to put a great deal more investment in it than we have seen in the past.

This is a very exciting time. There is a lot that needs to be done in the area of education and also in pushing beyond the misconceptions of what renewables are. Hydropower should reach its full potential. All of us, people in the administration, the new Congress, state governments, the environmental community in particular, and the industry can move forward and have a tremendous impact on the global warming agenda.

THE POTENTIAL FOR NUCLEAR POWER

Richard Wilson

Harvard University
Director, NE Regional Center of
National Institute for Global Environmental Change (NIGEC)

INTRODUCTION: THE TECHNOLOGY OF NUCLEAR FISSION

I have been asked to discuss the potential for nuclear power in the years ahead, because generating power from nuclear fission does not lead to emission of greenhouse gases; and therefore replacement of any fossil fuel electricity generating plant by a nuclear one will reduce emission of greenhouse gases. In a talk given to this group three years ago, (Wilson 1989) I showed that the reduction of greenhouse gas emissions by changing from fossil fuels to nuclear power, and the reduction of emission of greenhouse gases by improvement of end use efficiency (loosely called energy conservation) are independent of each other. Both can be partially effective. It is stupid to reject either because it will not do the whole job. If the effect of rising greenhouse gas concentrations is as bad as most scientists fear, both are necessary. In particular I took some leading energy supply projections, and showed how simple modifications could lead to more nuclear power and fewer greenhouse gas emissions than otherwise. (figures 1 and 2)

In order to understand the present position of nuclear fission power, it is important to understand a few features about the technology and how it differs from fossil fuel burning. I will take natural gas as an example for comparison, because it is natural gas that has been compared with nuclear energy in several recent societal decisions; yet natural gas is a greenhouse gas, and when burnt produces a greenhouse gas.

Natural gas is brought from the well to the user by pipeline to wherever the user wants it - his power plant or his kitchen stove. At the end of the pipe he/she can light a match and get instant heat at the right place. The convenience is almost as great as using electricity. The burner is simple and needs little maintenance. For electricity generation, the recent development of "combined cycle" burners for base load leads to a thermodynamic efficiency of 53% which is 50% higher than the average (35~) for all fossil fuel generators. Moreover, there is a possibility in the future for direct use in fuel cells with an even greater efficiency.

The amount of useful energy in one gram of uranium is 2 million times that in one gram of natural gas. This simplifies transportation of the fuel, but it makes the use of it more complex. While on an atomic scale, the nuclear fission is a simpler process than combustion, it is hard to exploit this simplicity in technical devices. Small nuclear fission reactors have

Global Enery Strategies: Living with Restricted Greenhouse Gas Emissions, Edited by J.C. White, Plenum Press, New York, 1994

been made that need no attention for months or years; the reactors in USSR satellites are examples. But because of the energy density there is a chance that something can go wrong and cause large problems. The main complexity and cost of nuclear power is in coping with this safety problem, whether by initial design, by responding to critics, or by "onerous" regulation.

Nuclear fuel is cheap, and it is plentiful even at present prices. The plentiful nature of the supply has not always been apparent; when nuclear energy was expanding rapidly world-wide in 1965 to 1975, it was feared that the uranium would soon become scarce. But a modification of the technology with a breeder reactor, will enable an almost unlimited amount

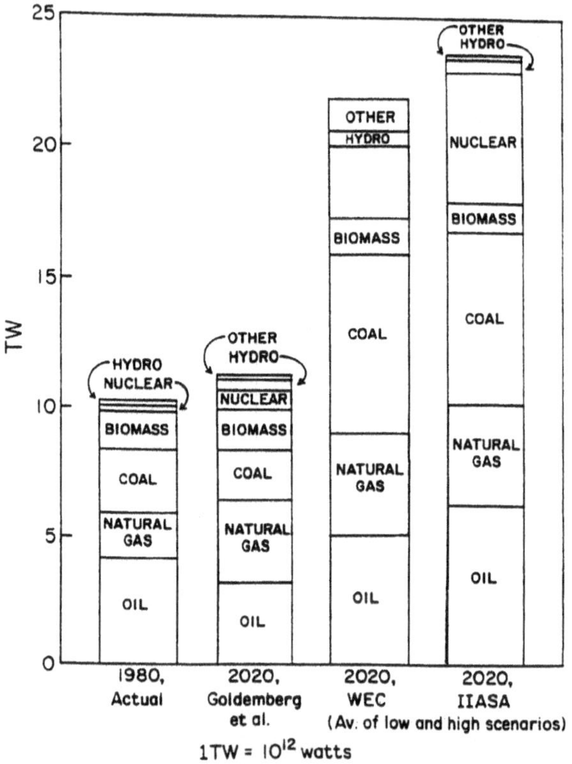

FIGURE 1. SOME ENERGY PROJECTIONS/COMPARED

of fuel to be available at an affordable cost; Many people have estimated for example that we can count on 100,000 years supply at the present world energy consumption using a breeder reactor. (Wilson 1972) Present estimates are, however that it will not be needed before the year 2020 and maybe not then. In this paper therefore I will not discuss the fuel supply in any detail. Nuclear fuel has a considerable advantage over fossil fuels; it is cheap to transport the limited amount of fuel actually needed, so that, provided that nations engage in international trade, every nation that wishes has equal access to nuclear fuel.

Natural gas is also plentiful. The world reserves of 120 trillion cubic feet has trebled in the last 10 years and are enough for 60 years consumption at today's rates.

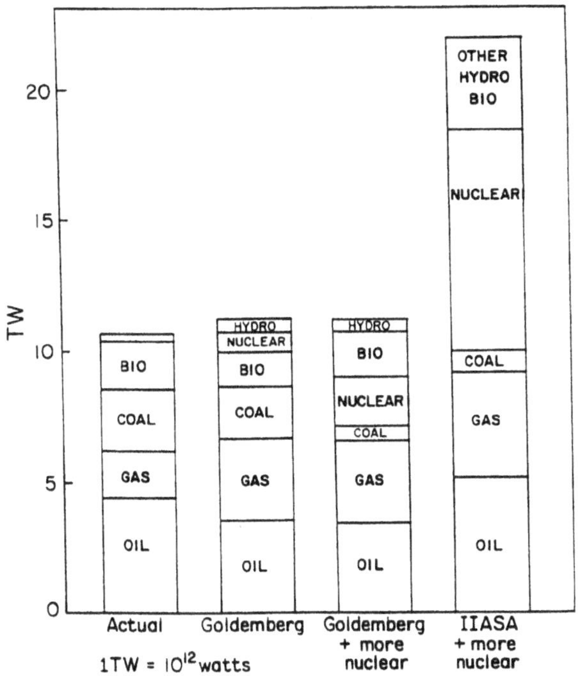

FIGURE 2. AS 1 BUT WITH MY SUGGESTIONS OF NUCLEAR POWER INCREASE

THE COST OF NUCLEAR ENERGY

Nuclear energy was once cheap, and competitive with coal, oil and gas generation. Now it seems to be expensive. In order to understand what the cost might be in the future, it is therefore important to understand what has changed and why it has changed.
Following Eisenhower's "Atoms for Peace" speech in 1953, there was euphoria about nuclear energy. It seemed to offer an unlimited, environmentally benign, source of energy to pull mankind out of poverty for ever. In 1970 it was also cheap; the busbar cost was 0.55 c/kwh from Connecticut Yankee, and 0.828 c/kwh from Yankee Rowe, although as noted below there was some federal subsidy for construction By 1980 this had all changed. Some environmentalists were actively opposing nuclear energy, and costs had escalated. This escalation has continued for the last 12 years. What had happened? Can we return either to the enthusiasm or the low cost? Should we try to return? I will spend most of the talk discussing these questions, and illustrating alternative courses for society, particularly U.S. society.

CAPITAL COST

The facts of the increases in cost are well documented, but not the reasons therefor. In 1961 Yankee Rowe cost $40 million to build for an initial 120 MWe installed capacity which was expanded to 185 Mwe. In 1966 Connecticut Yankee cost $120,000,000 to $160,000,000 (depending upon one's estimate of the value of government subsidies) for 550 MWe ($220 to $290 per kwe installed). In 1972 Maine Yankee cost $200,000,000 for $800 MWe also

$250 per kw installed. But this cost was increased 10% when Maine Yankee was forced to spend an extra $20 million to revise the cooling water system to meet environmental objections that had been raised after the initial design had passed the construction public hearing. This increase was just the beginning. Now even the best plants cost $2000 per kwe installed. This increase far exceeds the threefold inflation since 1972. The operating costs have also increased so that whereas in 1970 nuclear produced electricity was competitive with that from oil and coal, in 1990 it had become slightly more expensive; even though the cost of using oil and coal also increased in this period.

This runs counter to all previous experience. One expects that costs will come down as the new technology is learnt! The costs of most technologies have followed a "learning curve"; with nuclear power we seem to have a "forgetting curve"! A learning curve is evident in subsets of the nuclear data; the later nuclear power plants built by Duke Power cost less than the earlier ones. But superimposed is an overall societal increase of cost.

What had happened? A part of the problem is a general increase in construction costs. But unless the utility company accountants were lying on a massive scale in 1972, I can see only one main reason; a changed perception of the need for expenditure on safety, which is the main determinant of cost.

Some people claim that utility companies routinely added equipment and personnel when pressured by Public Utility Commissions, the Nuclear Regulatory Commission and environmental advocates without regard to cost. I am unable to contradict them. It is unclear whether increased cost led to improved safety.

A part of the increase in construction cost is due to interest charged on capital during the construction period. Interest rates have increased because of inflation since 1970; total interest charges have also increased because of delays. The delays in turn have been due in part to increased licensing requirements, (yet some older plants have had retrofit and the cost of those does not make up the difference in cost) a part is from public opposition, and a part may be due to construction by less competent utility companies. There is a wide variation in these cost increases, sometimes, but not always, associated with public opposition. A part, but not all, of this cost increase has appeared in other non-nuclear construction projects. A proposal 8 years ago that the National Academy of Sciences study this question came to naught. Those closest to industry seemed not to want to know the answers, because in many cases it could be embarrassing. Since we do not know for certain the reasons for this increase in cost, it is hard to predict when or even whether the costs will decrease again.

Following the accident at Three Mile Island in 1978, the staff of the Nuclear Regulatory Commission were insisting on design changes (in the name of safety) that no other industry has had to suffer; in some cases opposition that arose after the initial order (with its' construction permit hearing) has raised issues already technically decided and prevented operation after construction. delays were caused by protracted hearings before the Atomic Safety and Licensing Boards (ASLB) with very little substance. These changes and delays, led to unanticipated increases in construction cost. Public Utility Commissions hold "prudency hearings" to decide whether any particular expenditure was prudent. While it seems reasonable that utility companies be prudent, this concept is not evenly applied. For capital construction public utility commissions have declared that the utility company should have been able to avoid cost overruns, including some caused by opposition or excessive regulatory requirements. They have refused to allow the utilities to recover their investment, in decisions that are almost unique to the nuclear industry (Wells, 1989), (NAS, 1992).

In contrast to the effect of prudency hearings in discouraging construction, the "fuel adjustment charge", which becomes an ever larger part of my electricity bill, ensures that even if the utility company was imprudent and projected a fuel cost for oil or gas that is too small, they can recover <u>all</u> of the increase from the customer. With gas they are even better off. In the early 1980s many gas distribution companies signed "take or pay" contracts with suppliers. Prices were high and shortages were threatened; the courts have allowed the companies to walk out on their contracts!

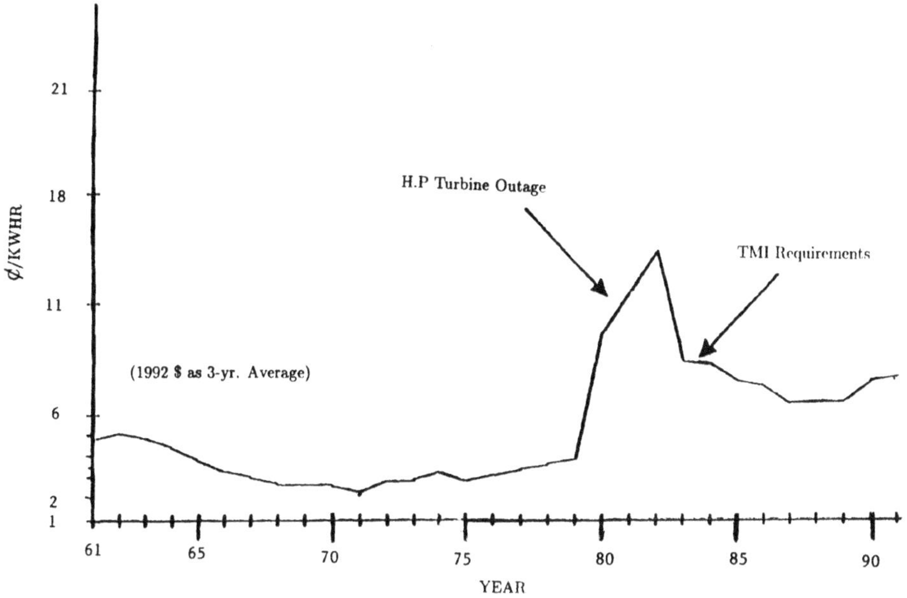

FIGURE 3. TOTAL COST OF YANKEE ROWE

OPERATING COSTS

The increase in capital cost for a nuclear power plant discussed above became very evident in the 1980s and has received a lot of attention. But far more insidious has been a steady increase in "Operations and Maintenance" (O and M) costs. Leading nuclear scientists told the nuclear industry at the beginning of this last decade that "if you operate the nuclear power plants safely for the next 20 years, all will be well". They were overly optimistic and ignored the effect of increasing costs. Several events in the last year bring to our attention the effects of this ignoring of the operating costs. In Figure 3 I show how the total cost of operating Yankee Rowe has changed over the years. Before 1970 the cost was mainly pay back of the "loan" or charge against capital. The 1970 cost of 0.82c/kwh is equivalent to a little over 2c/kwh in the 1992 dollars plotted here. The charge for construction cost must have gone down; utility company practice has been to charge the construction cost early; moreover inflation must have diluted the payments. But the operating costs went up; (Figure 4) this has been due to safety improvements demanded by NRC and also to an increase in Operations and Maintenance. A clue comes in the plot (Figure 5) of the staffing of the plant. The number of employees went up threefold over this time. I do not have a further breakdown but it has been claimed that a large part of the increase in employees was due to increase in the number of security guards (which may or may not have been accompanied by an increase in security). Here the treatment of energy sources is unequal. My local LNG facility has few, if any, guards till a shipment comes, and the 500 MWe hydroelectric (Comerford) dam is completely unguarded. I am able to drive my car out upon it unnoticed, and could lower a 1000 lb bomb over the side. Yet this much modern explosive could rupture the dam and carry away all the dams downstream, and many communities along the Connecticut river. This is not an idle speculation. In 1944 a hydroelectric dam in Germany was destroyed by a single 500 lb bomb; placed in position under unfavorable circumstances. It was dropped from the air

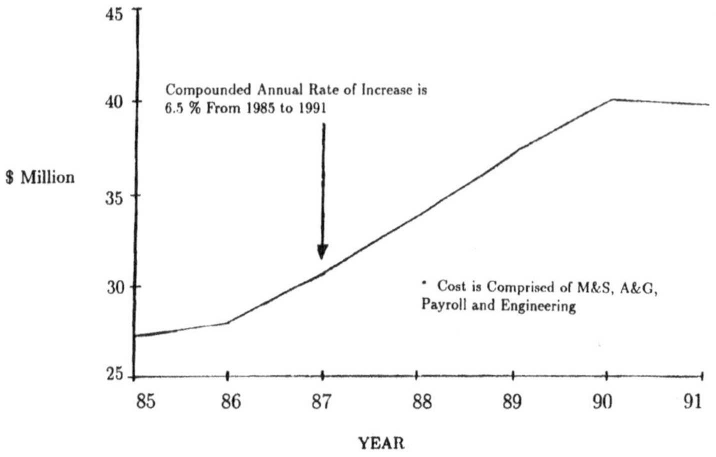

FIGURE 4. OPERATING COST OF YANKEE ROWE

against intense anti-aircraft fire and the best nets to stop torpedoes and bombs that Germany could devise. How much easier it would be with no opposition! In 1975 a study of California dams made by a group of scientists in UCLA (Ayyaswami 1974) showed that there are no evacuation plans, and in the event of a severe earthquake, many thousands of people could be killed.

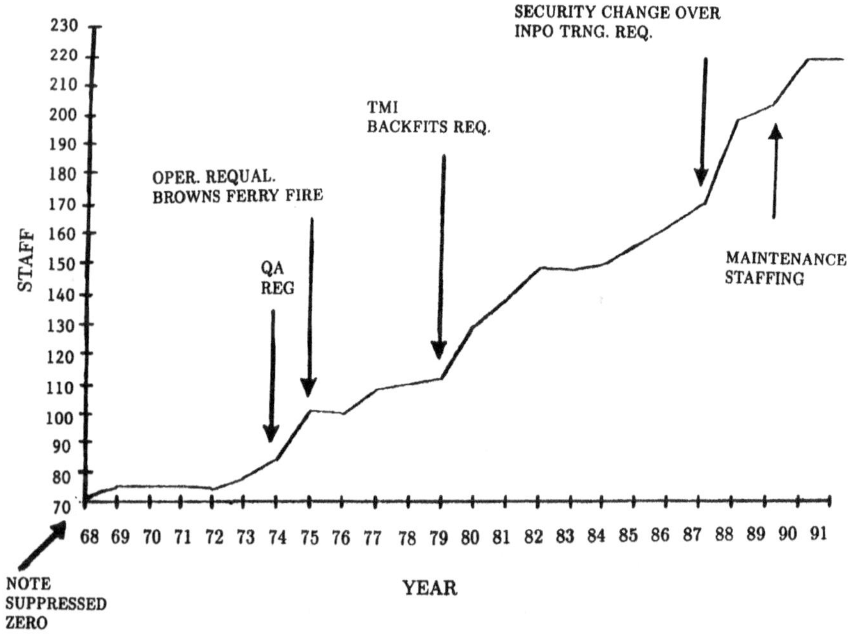

FIGURE 5. STAFF OF YANKEE ROWE

In Figure 6 I show an average of O and M costs for the industry as a whole. Even over the last few years, O and M costs have increased at over 3% a year! Although they seem to be flattening off, it would be a bold man who will say that O and M costs will not increase again. A part has been due to pressure tube failures in Westinghouse's steam generators. A close examination shows considerable variation among plants, suggesting that there are technical matters, such as corrosion that can contribute in many cases (Hansen et. al. 1979). This variation between plants has also been emphasized by Mahoney (1992). In my introduction I stated that much of the cost of nuclear power was because of the necessity of a safe controlled reaction. It follows then that much of this increase of O an M costs is in some way related to safety; whether from an increased industry perception of the need for safety, whether from direct responses to regulation or a decrease in efficiency of addressing safety concerns. It also follows from the variation in such costs among plants that some plants are likely to be more expensive than those using alternate fuels and are vulnerable to attack on this ground.

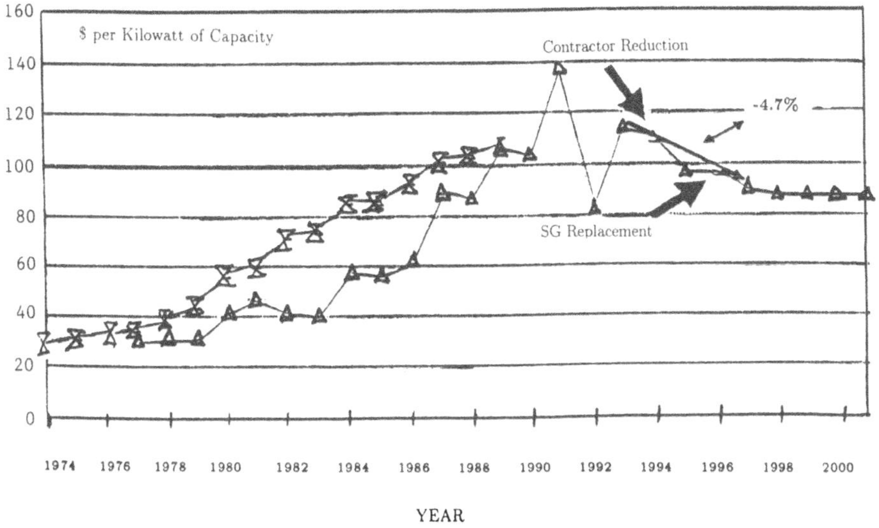

FIGURE 6. OPERATING FOR INDUSTRY (FROM TROJAN)

THE ANTINUCLEAR STRATEGY

Already in 1970 the nuclear euphoria of 1953 was not universal. Other views began to be expressed. Various scientists, including Dr Ernest Sternglass, Dr John Gofman, Dr Thomas Mancuso, and Dr Karl Morgan had already attacked atomic bombs and exaggerated the effects of radiation on man in order to do so. At the meeting of the American Association for Advancement in Science the then President, Nobel Laureate Glenn Seaborg was picketed. Not for his part in making the atomic bomb, or his work as Chairman of the AEC in assisting in Kennedy's build up of nuclear weapons, but because of his espousal of nuclear electric power. Professional societies, with a notable exception of a strong disagreement with the expressed views of Dr Sternglass by several past Presidents of the Health Physics Society, were silent. The public organizations that engage in research on the effects of radiation, the National Cancer Institute (NCI), the National Council on Radiological Protection (NCRP), and even the International Commission on Radiological Protection (ICRP), did little to contain the hysteria. The scientific and technical community were, and still are, silent in spite of an eloquent appeal by an English health physicist Dr Rotblat. This left the field wide open to extremists who were willing to distort the truth. Too few scientists were, and are, willing to speak up in public for scientific truth and process. Lay people therefore rally to the side that is open and enthusiastic.

By 1975, antinuclear activists had begun their steady, and presently successful in the USA, attacks. It is instructive to understand their methods. Although the public hearing process for individual power plants leaves more opportunity for intervention than for other power plants, it is continuously attacked as being not open enough. Studies made by government, industry, academia and non profit groups continually show that nuclear power is more benign than coal or oil burning (IAEA 1991). This led Ralph Nader 15 years ago to propose his successful strategy of using delays in the legal system to make nuclear power too expensive; this included the strategy of controlling the local public utility commissions. As Nader said early on: "We may lose every battle in the hearings, but we will win the war." The US legal system is particularly suited to such tactics. Few, if any, courts are willing to admit that delay, in itself, can deprive people of their legal rights. Yet justice delayed is justice denied.

THE DIRECTOR'S DILEMMA

The success of the US antinuclear power movement by 1983 is apparent in several ways. One may be summarized in the oft quoted "Director's Dilemma" One imagines a Director of a utility company who is convinced that, in the long run nuclear power is:

- cheaper than all alternatives
- environmentally superior to all alternatives
- a better neighbor than alternative power plants
- in the public interest.

Nonetheless he will not order a power plant unless he knows:

- what the power plant will consist of when he orders it
- what the power plant will cost to build and operate
- that he will be allowed to complete it when he has ordered it
- that when it is finished he will be allowed to operate it
- that when he operates it he will be allowed to recover his investment.

These seem like simple and obvious requirements, which are met in almost all industries. In 1970 a utility executive knew all of these things, or thought he did. By 1980 he knew none of them.

All of these requirements, and the costs, depend critically upon the political situation which in turn is associated with the fact that many members of the public do not understand and consequently fear this new technology. They do not know, and scientists forget to remind them, at the simple scientific fact about nuclear fission that lead to a huge differences in technological possibilities between nuclear and fossil fuels. The energy density is 3 million times as great; it takes three million times the weight of coal to produce a certain amount of energy as of uranium 235. This difference is the difference between chemical and nuclear energy densities. This difference enables mankind to make bombs a million times more powerful than before. Whereas in the second world war, a 50 pound bomb often destroyed a house, and a "blockbuster" had 2 tons of TNT, we glibly talk about bombs with a hundred million times the explosive power. Many scientists believe that the connection with bombs is the most important impediment to nuclear power. It is, however, important to realize that these bombs can be, have been, and probably will be made whether or not the world decides to use nuclear fission for peaceful purposes.

The difference in energy density between nuclear and fossil technologies enables many environmental advantages to be achieved. The quantity of fuel is small enough that we can afford to chemically purify the uranium both before burning, and after burning: which is not possible for fossil fuel burning. Although the waste products are highly toxic, they can be kept concentrated and their volume kept small. The waste products from fossil fuel, particularly coal, burning are also toxic, and their volume is inevitably a million times larger. The total toxicity is comparable - initially somewhat more for nuclear, but somewhat less after the radioactivity decays. People are often confused; they correctly attribute to high level nuclear waste a high specific toxicity (toxicity per unit weight), but forget that there is much smaller quantity! Indeed, as I have noted many times, this means that nuclear waste is the only waste in society for which we have a reasonable solution! But this is not the general perception of the public. It is vital to realize that the concentration of the nuclear waste is an advantage - but an advantage that can be thrown away by an inappropriate public emphasis. The public should be emphasizing that the waste must be kept concentrated; and this advantage not negated by faulty handling such as at Hanford.

A committee set up by the Forum of Science and Society of the American Physical Society stated that: (Hebel 1978) "we anticipate no difficulty in locating several suitable sites in different geological media within the immediate future". Immediate appeared to mean before 1985! They emphasized that it is an institutional and political problem, in which technologists can help, rather than a technical one in which politicians can help. Such a realization could spur people to search for those peoples or places where it is politically desirable to accept nuclear waste. The tentative offer of the Peoples' Republic of China, in exchange for help in construction of nuclear power plants, to allow nuclear waste disposal in the Gobi desert (for a fee) may well be one of these possible solutions.

Scientists were aware of the advantages and disadvantages of nuclear fission as they returned from the second World War in 1945. The perceived advantages drove them to develop nuclear energy. For a while they achieved widespread public acceptance for their point of view. The Joint Committee of Atomic Energy of the US Congress ensured bipartisan political support. Now, however, many in the public are suspicious of scientists. The joint committee was disbanded and fifteen committees vie for the task of controlling the Nuclear Regulatory Commission, and criticizing its' activities. Moreover there is, world wide, an anti-scientific trend: described eloquently by Kapitsa (1991) .

In these circumstances, the fastest route to early retirement for a utility company president would be proposing a nuclear power plant. However, any one proposing a new gas plant, can be sure that he can make money, once the plant has been approved. I repeat what I said widely in 1975. We are not sure of the cost of nuclear electric power. It obviously can be as low as it was in 1972, and could probably (due to learning) be somewhat lower. But antinuclear activists can, if we let them, force the cost to rise without limit. This is one of the many situations in which I am sorry to be right.

THE ROLE OF THE STATES

The Atomic Energy Act preempts state legislation in the fields that it covers, particularly safety including radiation safety. However, many states have nibbled away at this. For example it is clear that California reserves the right for itself to determine the adequacy of any procedure for nuclear waste disposal. (California 1976). While accepting the principle of federal preemption, the US Supreme Court has accepted the constitutionality of California's law in a decision which puzzled many observers. Tribe (1983) called it "a total victory for the states". The best discussion of this whittling away of federal authority is given by Pasternak and Budnitz (1987). The state role remains preeminent in items of cost and price and is typically controlled through public utility commissions (PUCs) which are very sensitive to local political issues. The California waste disposal law was accepted by the Supreme Court as constitutional because it allowed the PUC to exercise *economic* control. It did not however, prevent the economic investment by state utilities in cross border power plants. (e.g

Palo Verde in Arizona). This decision reinforced the view of many who questioned the wisdom of nuclear power and realized early on that the states could prevent nuclear power whatever the federal government decides on safety.

I now illustrate the way in which PUCs can shut down complete and operable power plants by discussing several cases where this has happened. I do not possess enough detail on any of them to call them "case studies" in any formal sense; indeed enough information is hard to acquire. Therefore some of the views here expressed should be considered more as questions for others to answer than rigorously derived conclusions.

SHOREHAM

The Shoreham nuclear power plant of Long Island Lighting Company was proposed about 1972; interestingly enough, the request for licensing was within a few days of a request for a construction permit for a similar boiling water reactor at Millstone Point - just across Long Island Sound. There were delays in obtaining a construction permit at Shoreham; the application came a few days after a moratorium caused by the adverse Calvert Cliffs decision, (causing an 18 month delay) and there were costly mistakes in construction so that over $6 billion was finally spent on a plant that cost NE Utilities $425 million at Millstone Point! Although the engineers and the federal (NRC) regulators felt that the plant had successfully surmounted these hurdles and was safe, the loss of public confidence led to the demise of the plant. The county, who had originally supported the plant at Shoreham, changed their minds and opposed it. An opportunity came to block the plant when the NRC (1980) issued a new regulation which insisted that the local community approve the emergency plan as a condition of an operating license. Both the county and the state declined to approve the plan, and held up the operation of the plant. After several years delay, NRC prevented this effective "veto" by the local community and the state by modifying the regulation.

But in the intervening years the Governor and PUC made an offer Long Island Lighting Company (LILCO) could not refuse. The state would buy the plant for $1; they could declare the plant a "tax loss" and get back an appreciable fraction of the extraordinarily high cost of the plant against the federal tax bill, and the rest against electricity rates. Although the federal taxpayer paid a fraction of the original cost (inflated by inefficiencies and delays) they paid almost as much as the plant was worth! From the point of view of the LILCO ratepayers, this seemed ideal. The one scientist in Congress, Representative Don Ritter, tried to stop the allowance of the tax benefit, firstly by asking the IRS to rule it invalid and then by a special bill. It seemed to him that it was a terrible precedent to allow someone to abandon it as an imprudent investment when the plant could in fact work economically once sunk costs were ignored. Several citizens of Long Island, joined by the US Department of Energy, tried unsuccessfully to force the state to write an Environmental Impact Statement so that the environmental costs could be publicly presented and properly considered. (The local Shoreham school committee continued to support continued operation until a hole was drilled in the reactor vessel to ensure that no one could change their minds). It is noteworthy that the alternate source of electricity is the burning of oil or natural gas. In the cost comparisons no allowance was made for the environmental cost of emission of greenhouse gases.

YANKEE ROWE

I have mentioned before that Yankee Rowe produced cheap electricity in New England in 1970 at 0.95c/kwh (about 2 cents/kwh in 1992 dollars). But it was also an old plant. By 1991 the costs had risen to 7.1 cents/kwh. In summer 1991 concerns were raised by NRC staff and others. "Had the reactor vessel become dangerously brittle from neutron bombardment?" The NRC staff originally agreed with carefully argued presentation by Yankee Atomic Corporation, the owner of the power plant that it had not. But after intervention by a group critical of nuclear safety, the Union of Concerned Scientists, (UCS) who also claimed that Yankee Rowe was a lot more dangerous than more recent plants, this was reviewed. This review was encouraged by the new Chairman of the Commission, Dr Selin, who apparently wanted to make his name as a tough regulator. The plant was shut down while new tests were to be conducted. These new tests were to cost only $28 million (Kadak, 1992), (less than 0.1 cents/kwh amortized over 20 years) but the utility felt no confidence that the tests would actually satisfy the NRC, and that more would not be demanded - particularly at the time of license renewal in 2001. A calculation suggested that over the next ten years other sources of electricity would be cheaper. Although not stated it is likely that natural gas (or oil) will effectively be the replacement (although a new coal fired cogeneration plant of the same size as Yankee Rowe has just been approved for Eastern Massachussets)

I note that no allowance was made in the comparison for the environmental cost of the emission of greenhouse gases, and no one protested on behalf of environmental diversity or preservation of the habitat of Canadian Indians when New England proposed to buy hydropower from Hydro Quebec.

An interesting facet is that if Yankee Atomic, the owner of the plant, had spent the $28 million on studies and still not been allowed to operate again, the costs might be called imprudent and then the company might not collect from the ratepayers. If the studies had been made while the plant was operating, a different situation would have prevailed and they might have decided differently. This is an example of a general case; plants should not be shut down while studies are made unless there is a real safety emergency.

SAN ONOFRE I

San Onofre I (SONGSI) is a 500 MWe reactor which has been operational since 1968. It has just completed a record continuous run without shut down for maintenance. It comes to the attention of the California PUC when NRC insisted on $150 million in capital improvements (to make it more earthquake resistant) in 1980. The alternative considered was to shut down and use natural gas using a new combined cycle generating plant. Various scenarios can be made about future prices, but if it is assumed that O and M costs for nuclear power continue to rise, and natural gas prices do not rise for the next 6 years, (Figure 7) shut down becomes economic.

The California PUC had suggested that Southern California Edison run San Onofre I as an Independent Power Producer (IPP). They declined. Presumably they would sell the power plant for $1, and even make an adjustment for costs of decommissioning. If the PUC are wrong in their economic forecasts, anyone who buys and operates such a plant will make money. Where are the pronuclear millionaires who can afford to make such a gamble? Alas, it seems that the only pronuclear people left are a few crazy, starry-eyed academics with no money!

The California Energy Commission and the PUC staff had recommended that environmental issues be taken into account in such comparisons. However, no allowance was made by PUC for the greenhouse gases emitted by the replacement power plant. If it had been made, the balance would have thrown the decision the other way.

TROJAN

The Trojan nuclear power plant in Oregon has a larger capacity than either of these (1130

Mwe) and the economic factors might be thought to be superior. However the management of PGE have made a "final" decision to close Trojan at the end of its current fuel load in 1996 (Trojan, 1996). Again, however, the continued operation of Trojan was compared with new combined cycle gas generation and, on the assumption of a 1.5% yearly increase in O and M costs from 1993 to 1996 and 3% above 1996, leading to another doubling of costs by 2017, but little rise in gas prices, was found wanting.

These may seem unreasonable 0 and M projections. In a revealing comment PGE explicitly say "what was ultimately chosen was a compromise between the plant's input and other interested parties inputs" (Heintzmann, 1992).

Although Trojan claims externalities have been figured in, inadequate allowance has been made for the environmental costs of emitting greenhouse gases. It also appears that the

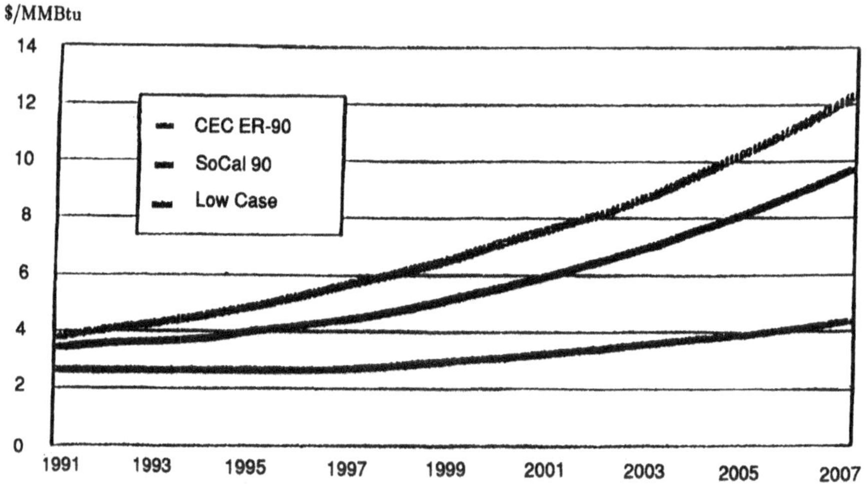

FIGURE 7. NATURAL GAS PRICES NEXT FEW YEARS

scenario assuming spot prices of natural gas has already been shown to be defective by the doubling of the spot price after the hurricane in September 1992 forced the temporary closure of several off-shore gas wells.

What allowance should be made for the emission of greenhouse gases? In a paper I presented at an earlier conference in this series I discussed this question. (Wilson 1988) Many economists have done so also. Economists discuss a "tax" on carbon of $40/ton (Nordhaus, 1991), (Jorgensen and Wilcoxen, 1990). If allowance is made for the different amounts of CO_2 produced, this will amount to 0.08c per cu ft of natural gas, or about 0.6c/kwh of electricity with 50% efficiency. This was proposed by Ross Perot during the presidential elections and it is not improbable that it will be imposed during the next ten years. Although 0.6c/kwh seems small it would turn the PUC decisions around in some cases.

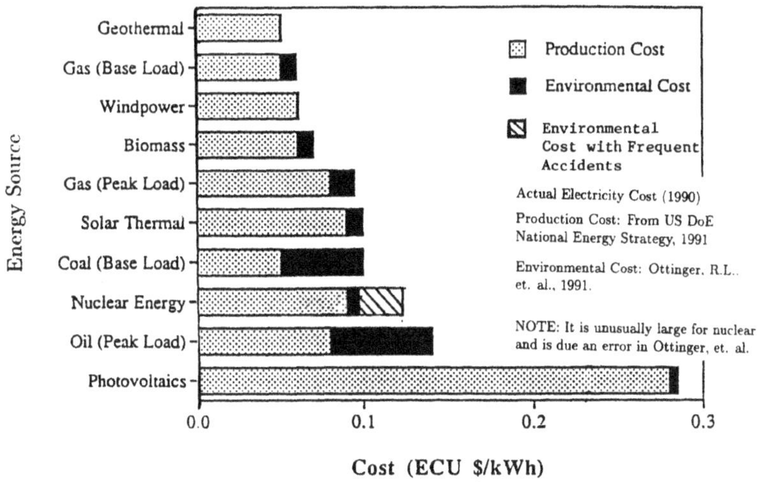

FIGURE 8. ACTUAL ELECTRICITY COST (1990)

In this PGE ignored an additional important point (Heintzmann 1992). CH4 is a greenhouse gas which is 23 times as important, molecule for molecule, as C02. Even if 4% of the gas leaks anywhere between the well and the power plant the effect (together with an assumed tax) is doubled to l.lc/kwh. An English study, (Grubb 1991) suggests that a 10% leakage rate is possible. Moreover sometimes CO2 comes out of the well with the gas. In one field in Indonesia, which may well be supplying California, four CO2 molecules come out for every CH4 molecule, multiplying the effect by a factor of 4! The Oregon Public Utility Commission (OPUC 1992) explicitly opted to ignore this also.

SHOULD WE RESURRECT THE NUCLEAR OPTION?

Is the success of this antinuclear strategy a great success of mankind over those who would misuse the forces of nature, or is a stupid refusal of mankind to understand the physical world in which we live, and to use God's bounty for the benefit of all mankind? It has been common for antinuclear activists to state that opposing nuclear power is not a technical but a moral issue, and that nuclear power is an intrusive evil. I suggest that while admitting that nuclear power may pose more moral issues than technical ones, it should be considered immoral to willfully oppose a technology that can improve the living standards of a number of the world's poor. Nobel Laureate Andrei Dmitreyvich Sakharov, speaking at the"Forum for a Nuclear Free World" in Moscow in February 1987, reproved his German "Green" colleagues. He suggested that instead of opposing nuclear energy, they work to make it safer; because the world will need nuclear energy as it strives to help the developing countries.

THE MIXING OF BOMBS AND POWER PLANTS

In 1946 nuclear physicists and others returning from the war, did not want the atomic bomb to be under military control, and insisted upon a civilian Atomic Energy Commission (AEC) to oversee all uses of nuclear fission. This decision, however useful it may have been in controlling military excesses, laid deep problems for peaceful uses. For many years, military uses, and military habits of secrecy, influenced the Commission. A myth arose that bombs and nuclear power stations are inseparable, even though most power station engineers know less than many bright undergraduates about how to make a bomb, and no nation has ever used a nuclear power program in the quest for nuclear weapons. This mixing has led to official secrecy and a confusion of thought eagerly exploited by a few antinuclear scientists.

There is no doubt that the system and the people who are knowledgeable about a nuclear fuel cycle can be used to plan and build a fuel cycle for bomb making. But such people can also prevent clandestine bomb making. This is a vital issue which needs far more discussion then I can give here.

In discussing nuclear safety, a committee of the International Atomic Energy Agency (REF) discussed two levels of accident probability; firstly a tolerable limit above which the technology should not proceed. The law, as interpreted by the US Supreme court In Silkwood (1984) clearly says that "the promotion of nuclear power is not to be accomplished at all costs". The Chernobyl accident clearly exceeded this. Secondly there is a "de minimis" limit where the accident is generally regarded as impossible, and needs no further thought. The Nuclear Regulatory Commission and the nuclear industry have been addressing the second, de minimis, level. But it could be argued that if the alternative is the closing of an acceptably safe nuclear power plant, should not the comparison be to the safety of the alternative technology calculated in a similar way? This is not done at the present time.

It is paradoxical that two states which have been leaders in urging "least cost" energy planning, New York and California, rejected a request from citizens and others, including the Council for Environmental Quality, that they prepare an Environmental Impact Statement for the proposed dismantling of Shoreham and Rancho Seco although the stated purpose of an environmental impact statement is similar to the least cost energy plan. New York and California fought the requests in court (and won). This has led to speculation that the call by these PUCs for consideration of environmental factors is insincere; as presently applied they must be mentioned, but are not included in any decision. Would the present uses of fossil fuels satisfy the published "safety goals" of the Nuclear Regulatory Commission if an appropriately conservative view is taken of the effects of air pollution, or of the likelihood of war as we squabble over the price of oil?

Even natural gas, which is the particular fuel in the comparisons above, is not completely benign. There is a long list of risky locations; from accidents in drilling; fires on off-shore rigs; pipeline explosions, and explosions at the end user. Supermarkets have been destroyed in the dead of night; and if the contents of a typical LNG tank were mixed stoichiometrically with air, and ignited, the explosion would be the size of 20 Hiroshima bombs. Colgate's (1974) scenario of the hazard of natural gas getting into sewers, by evil intent or otherwise, was laughed at. It is similar to an accident that occurred in Mexico recently and has never been properly considered. Bad operation, similar to that at Chernobyl, caused an accident in the trans-siberian pipeline in 1987 which incinerated 300 passengers on a passing train. These and other accidents happened with Liquefied Petroleum Gas (LPG) which is heavier than air. It is likely that for LNG which is lighter than air except when in large quantities in a cold cloud, the accident probability is much less. Nonetheless, this is rarely argued.

Public Utility Commissions often now demand "Least Cost" energy planning. While it is far from clear that the procedures specified lead to a lower cost than the procedures that they replaced, "least cost" plans are supposed to ensure that all factors are considered including environmental factors. But in all the cases above the main environmental concern about natural gas, the potential increase of the greenhouse effect, was ignored or incorrectly calculated.

Comparisons of environmental hazards have been numerous in the last 20 years. I note in particular three studies: by a French "colloque" (SFDN 1980) by Ottinger (1991) and by an expert symposium of a dozen UN agencies (IAEA 1991), Ottinger's study of nuclear energy costs shown in Figure 8 assumes a Chernobyl-type nuclear accident every few decades

- an assumption that no expert would make. However, I include it here as an example of a study that corresponds to public perceptions; that has been used, and will continue to be used unless decisively contradicted. Both the French "colloque" and the UN study puts nuclear power and natural gas on a par as regards environmental effects, and both much safer than use of coal or oil.

My tentative conclusion, however, is that if it were not for global warming, no one would worry about the replacement of nuclear power by natural gas. But in this conference we are explicitly considering greenhouse gases so we must worry.

OTHER DEVELOPED COUNTRIES

For completeness I mention, but do not elaborate, the situation in other countries. England, Germany and Sweden seem to have a situation not unlike that in the USA. Austria, Denmark and Italy have abandoned nuclear energy. France alone of European countries has a well organized plan for construction, operation and paying for nuclear power plants which is of some envy in the US. Although their success is often attributed to having one type of reactor, the latest in the series differs considerably from the earliest. I contend that it is their planning which distinguishes them. Japan has many diverse reactors; but although it does not produce as much nuclear electricity as France, it also has a well organized plan.

The economic and political situation in eastern Europe is confused. Electricity is sold very cheaply - at less than half the fossil fuel price on the international market. The incentive, therefore, to use nuclear energy where the labor costs are internal, or payments can at least be made within the former COMECON system is considerable. They are trying to continue and expand nuclear energy.

Western doubts about safety of Soviet designed reactors led many western countries two years ago to call for their shutdown - a plea echoed by some of the East European people. However, there was unanimous agreement at a special meeting on safety of Soviet reactors that the proper course is to help very competent professionals in these countries upgrade their safety standards (ANS 1992).

DEVELOPING COUNTRIES

Every time a nuclear power plant is built instead of a fossil fuel power plant it will play its proportionate part in reducing the emission of greenhouse gases and limiting global warming. But there has always been an additional concern about developing countries. Will a developing country have a system like that in the USSR (which led to Chernobyl) or one like that in France? (which has had no major accident) It is common, particularly for people who call themselves liberals, to be paternalistic and to state that the technology is too hard for a developing country. But let us look at the record. In 1955 most westerners, arrogant as we are, would have considered Korea a developing country. Yet Korea has built up a nuclear power program which appears to be operated as well and as safely as any in the world. They have accepted the advice, help and training of the western world without developing an inferiority complex. The same applies to Korean industrial development generally. In contrast, Iran had a better start than Korea; it had oil money and a long intellectual tradition. But its political problems in the late 70s led it to abandon nuclear energy, and other industrial development has lagged. I was Chairman of a committee reviewing the operation of the nuclear power plants in Taiwan (Wilson 1992). It was abundantly clear to us that the Chinese are technically, and in Taiwan politically, as capable of running these plants as anyone in the world. Since our report, the Republic of China had decided to build a fourth nuclear reactor unit of two reactors by the year 2000.

In 1982 I visited Egypt and discussed Egypt's hopes for a nuclear power plants to be built at El-Dabah, near the large populated area of the Nile Delta. An elderly engineering Professor asked whether I thought that Egypt was capable of operating such a plant. I reminded him that in 1956 the British seriously stated that Egypt was not competent to run the Suez canal by themselves. Yet after Egypt took it over from the British Government, many needed modernizations and improvements were made without fuss and fanfare; the canal has probably been better and more safely run than before. But it will be hard to compete with fossil fuels. Oil and natural gas are sold internally at prices far below the international price, making an effective subsidy. If capital can be provided to provide an equivalent subsidy for nuclear energy Egypt seems a good candidate for expansion.

The fundamental feature of nuclear power, its energy density, helps developing countries. To run a successful and safe nuclear power program it is not necessary that all the people in the country have technical training. I note that an attempt to develop a windmill program in Egypt in the 1960s had failed; there were too few technicians for maintenance. But for nuclear energy one merely needs a small number of highly technically trained people. As countries develop, this happens naturally as the bright and privileged few are educated overseas. This can be called, and is, elitism. It may be an unpopular thought to the starry eyed liberals of Berkeley (CA) and Cambridge (MA). But is it wrong? I suggest here that those developing countries that can politically accept a technically trained elite, can have a successful nuclear power program. But this technical training inevitably is associated with a degree of political freedom that is sometimes difficult for politicians of a developing country to accept. The Philippines have built a nuclear reactor. Politically we did not help; allowing domestic opposition to nuclear power to delay shipping of parts to the Philippines. Worse still, the reactor was associated with the old, discredited Marcos regime. There is a widespread belief that Westinghouse paid a large bribe to Marcos' family. It has taken several years for the perception of corruption to fade. Only now are steps being taken to bring the power plant into operation. But this could be a promising new beginning.

Can we allow the "free market" between nations to decide which countries are "allowed" to have nuclear power? I think not. All the world must worry about a country that tries to develop nuclear power but fails to support the technical elite; they will have power plants that are badly run; there will inevitably be political pressures to cut corners to provide output at the expense of safety. Already Europeans and Americans see the importance of ensuring that the power plants in the former USSR are run safely. We do not want another Chernobyl. How the world is going to give this help without being charged with interference in domestic affairs is an interesting challenge.

The present situation of the republic of Armenia is instructive. 3 years ago, after the earthquake, there was public pressure to shut down their two W ER 400 reactors. They were built in an area of high population density, and in an earthquake zone. Three US engineers of Armenian ancestry visited the plants and reported (Hadjian 1978) that although the reactors were earthquake resistant, the auxiliary systems were not. The central government of the USSR shut the plants down. Now Azerbaijan has cut off oil supplies, and intercepted the natural gas pipeline from Russia. Armenian industry is at a standstill. There is now a considerable movement to restart them, even though there will have to be extensive rehabilitation. The new President of the republic, Lev TerPetrossian, was among those who 3 years ago wanted the plants to shut down; but now the desperate situation of his country forces a change in thinking. Armenia is searching for international sources of capital for a loan to recommission the plants. This example shows that political opposition is likely to reverse itself when other more troublesome political situations dominate. It seems vital, therefore, to ensure that neither San Onofre I or Trojan be actually destroyed in the way Shoreham has been.

It is useful to realize that the low transportation costs of nuclear fuel make nuclear energy particularly competitive in countries with no indigenous fuel supplies such as Japan and

Taiwan. Such countries will also find that an attractive feature of nuclear energy is the ability to store many years of supply on site: compared with a typical 3 months for a fossil fuel power plant. This leads to a degree of political security that may match the historical needs of the country. Japanese are fond of reminding us that they entered the first world war on the side of the British (against the Germans) to safeguard supplies of Manchurian coal; and entered the 2nd world war against the British, Dutch and Americans, because of the oil embargo against them of summer 1941.

TECHNOLOGICAL IMPROVEMENTS

Recently much fuss has been raised about a "new generation" of reactors that is safer than the old. I believe that reducing this probability further will have no influence on the anti-nuclear community. Indeed a risk 1 benefit analysis gives a negative balance, however low the risk, if the benefit is zero or negative as many anti-nuclear people perceive. The calculated probability of core failure is 10^{-5} instead of 10^{-4} (REF). The discussion above leads me to claim that the important issue is whether this safety advantage can be translated into a lower construction cost, and even more importantly a lower operating cost. One way might be less regulation. If this can be done, nuclear power may revive. Since an outsider cannot easily tell who is responsible for which cost, all I can do is recommend intensive thought by the utility industry and by the Nuclear Regulatory Commission.

There has been an order from the NRC commission to the staff to make rule changes only when these are cost effective. Yet the orders to Yankee-Rowe and San Onofre that forced their demise were based on earlier (vague) rules. Nor has this order yet shown itself in a reduction in O and M costs. Nor has it shown itself in a general belief that the cessation of increase is permanent.

The next generation of plants may be more standardized, so that the Utility Director may know what he is ordering. But whether that will in fact happen and lead to the needed reduction of cost is open to question. History has few recorded cases of relaxed regulation. C. Northcote Parkinson (Parkinson 1950) reminded us that the staff of the British Admiralty expanded between 1914 and 1928 by 74% even as the number of capital ships decreased by 68% and the numbers of sailors decreased by 31%. This increase in the bureaucracy comes to a little over 5% per year. Likewise, unless we take forceful action, the NRC staff will expand at 5% per year well after the last nuclear power plant is shut down!

In 1975, when the AEC was broken up, and ERDA and NRC were created from the wreckage, I proposed an "Energy Regulatory Commission" omitting the word nuclear. Maybe it is time to consider this; then all energy supply methods might be considered on an even basis. Then one can find out whether treated equally, nuclear power is expensive or not. If that cannot be done, maybe state legislators can enact a nuclear set-aside for operating plants to make sure that this option does not completely vanish.

Nuclear fusion has been studied for over 40 years as a source of electricity which is environmentally superior to fission, and offers an unlimited amount of fuel from sea water. It has suffered in the past from too much optimism. This led to a public discounting of any projection and has obscured the strides made in the last 15 years. A test reactor (Joint European Torus, or JET in Culham, England) has been close to achieving "break even"; generating more heat than the electricity consumed by the reactor itself. Since the fusion reactor is more complex than a fission reactor, the cost of a fusion electricity generator is likely to be higher than fission electricity generators if other factors are equal. But if the safety advantage can be translated into simpler regulation, fewer security guards and similar cost cutting mechanisms, nuclear fusion may well have a bright future. In 1991 a special advisory committee of the US Department of Energy recommended a program that might enable US to build an economically competitive fusion power plant by 2020 (Stever 1991).

CONCLUSIONS

None of the factors that have led to the increase in cost of nuclear energy, and other undesirable consequences, have changed appreciably in the last 10 years (in the USA).

- the scientific future for nuclear power remains excellent
- the technical future for safe nuclear power is steadily improving
- the economic future for nuclear power is still getting worse
- the political future for nuclear power still looks bleak.

If it is desired to continue nuclear power development in spite of the evident existence of some adamant opposition it is necessary to reduce the possibility that the opposition can cause expensive cost increases by legal delays. The recent "one-stop" licensing must be seen in this light. However federal action is not enough; the evacuation planning rule of 1980 gives a veto to the Governor of each state, even though since 1989 it has been only temporary and therefore merely a delay. It supplements the considerable state power of economic regulation. Even individual states have created and can continue to create a climate that prevents any resurgence of nuclear power in the USA.

I believe that the only hope for a resurrection of nuclear power in the USA is for a massive effort on public education by the scientific and technical community. Scientists must speak out. In particular, I call on physicists which field were the first proponents of nuclear electric power, to make public statements. They should do so soon before the industry has disappeared.

Some of the legislative and regulatory preconditions are in place. We must insist that they be used. The NRC has a sound set of "Safety Goals", but that does not prevent them proposing additions to make the safety projections exceed these goals. NRC in 1975 adopted a "definition" of "As Low As Reasonably Achievable" ALARA, that improvements be made if they cost less than $1000/manRem. Yet this does not stop NRC and EPA demanding expenditures (especially for waste disposal) that far exceed these amounts. Lay people must be asked to join scientists in demanding a proper comparison with other sources of electricity generation and appropriate changes in all aspects of the industry including its regulation. If nuclear proponents are correct these will enable economic operation again. Unless this happens soon, the present competent people in the nuclear industry will leave and new students will not be attracted. It will then be more expensive and less safe to start again. Although I do not now foresee such a change, it might happen fast.

For the countries of the Pacific Rim, Japan, Korea, Taiwan and even mainland China:

- the scientific future is the same as in the USA;
- the technical future is the same as in the USA;
- the economic future is bright and;
- the political future looks excellent.

This might be seen as one more example of why the next century will be an "oriental century." The resurgence of nuclear power may come from the orient; let us also hope that if and when it is again economically and environmentally attractive for the USA, our country will follow close behind. Otherwise our economy will inevitably decline and we will become an undeveloping country.

REFERENCES

ANS (1992) "Workshop on Safety of Soviet Designed Nuclear Power Plants", American Nuclear Society, Chicago, IL.

Ayyaswami P., B Haas, T Hsieh, A. Moscati, T.E.Hicks, D.Okrent (1974) "Estimates of the risks associated with dam failure" University of California at Los Angeles, UCLA ENG-7423 March.

California (1976) #25524.2 of the California Public Resources Code.
Colgate S. (1974) Privately circulated memorandum.
Grubb. M.J., (1991) Energy Policies and the Greenhouse Effect, Dartmouth Publishing Company, Hants, England.
Hadjian A.,et. al. (1978) Report to the Government of the Republic of Armenia.
Hebel, L.C., E.L.Christensen, F.A.Donath, W.E Falconer, L.J.Lidofsky, E.J.Moniz, T.H.Moss, R.L. Pigford, G.J.Rochlin, R.H.Silsbee, M.E Wrenn, (1978) "Report to the American Physical Society by the study group on nuclear fuel cycles and waste management" Rev. Mod Phys. 50 No. 1, part II Sl to S186.
Heintzman D.W., Director, Corp. Cummunications PGE, letter to R. Wilson.
Hansen, K. et. al. (1989) Technology Review.
IAEA (1991), "Senior Expert Symposium on Electricity and the Environment", Helsinki 13-17 May 1991. Published by International Atomic Energy Agency, Vienna.
Jorgensen, DW and Wilcoxen PJ (1990), "The Cost of Controlling the Greenhouse Emissions", Workshop on Modeling for Climate Policy Analysis, Washington, DC, (October).
Kadak, A (1992), Yankee Atomic Electric Company, letters to R. Wilson.
Kapitsa, S.P., Scientific American, October (1991).
Mahoney S. (1992) "PLEX: Nuclear Power Plant Life Extension or Extinction?", Public Utilities Fortnightly, Nov. 15.
Nordhaus WD (1991) "To Slow or Not to Slow: The Economics of the Greenhouse Effect", Economic J. (July).
NRC (1980) "Emergency Planning: Final Rule", US Nuclear Regulatory Commission; changes to 10CFR50, Federal Register 45, 55402 August 19th.
0PUC (1992), "External Cost Proceedings", Oregon Public Utility Commission (Docket UM-424).
Ottinger R.L. (1991) "Environmental Costs of Electricity". report From Pace Institute to U.S. DoE. See also criticisms of this report to DoE.
Parkinson C.N. (1950) ,"Parkinson's Law".
Pasternak, A.D. and R. Budnitz, (1987) "State-Fededral Interactions in Nuclear Regulation", Lawrence Livermore Laboratory, Livermore, CA, UCRL 21090.
SFDN (1980) "Colloque sur Les Risques Sanitaires des Differentes Energies", Societe Francaise d'Energie Nucleaire, Gedim, Paris.
Silkwood (1984), Bill. M. Silkwood,... appellant v. Kerr-McGee Corp... et.al. No 81-2159. Decision of the U.S.Supreme Court 464 US 238, 78L Ed 2d 443,104 S Ct 615. January 11th.
Stever, G.G. et.al. (1991) "Report of Fusion Power Advisory Committee (FPAC)" US DOE.
Tribe, L. (1983) Counsel for the California Energy Commission; quoted in Sacramento Union, April 21st.
Trojan (1992) "Least Cost Analysis for Trojan".
Wells CW (1989), "Prudence Audits Are Narrowing Our Energy Choices", Public Utilities Fortnightly, p. 11, May 11th.
Wilson, Richard (1972), "Kilowatt Deaths", Letter to Physics Today Vol 25, p.
Wilson, Richard, et. al. (1992) "Nuclear Power Operations in Taiwan", Report to the Honorable Vincent Siew, Minister of Economic Affairs, March 29th. (made public by the Minister in English and in Chinese).
Wilson, R, (1989) "Global Energy Use: a quantitative analysis" in Global Climate Change Linkages Ed J.C.White, Elsevier.

THE POTENTIAL FOR REDUCING CARBON EMISSIONS THROUGH EFFICIENCY - INTRODUCTION

Christopher Flavin

World Watch Institute

It is generally recognized now, not only in the analytical community but in the energy policy-making community, that the most significant single change that we have seen in global energy systems over the past two decades is the fairly substantial improvement in energy efficiency. This has allowed virtually all of the industrial countries to reduce the amount of money that they must commit to providing basic energy services. It has allowed many countries, including the United States, to significantly reduce their oil imports and, of course, energy efficiency has also allowed us to greatly reduce the amount of air emissions and other environmental impacts that result from virtually all of our energy supply systems.

Back in the mid-1970s, when Amory Lovins first introduced the idea of a soft energy path, his concepts and projections were taken as being the wild-eyed ranting of somebody who is really not fully balanced in intellectual facilities. Since then, if you look at the projections that Lovins made in 1975, you find that we are actually using less energy than he forecasted under his low energy scenario, which was dismissed out-of-hand by everyone from the Edison Electric Institute to the U.S. Government as being a nutty kind of a scenario.

One of the things that is probably not all that clear in everyone's mind, but which is central to the analysis that Amory Lovins originally put forth, is a distinction between energy efficiency and energy conservation. I am sure that it is important that you have a sense of what that difference is.

Energy conservation basically means the efforts we put forth to reduce energy use in a whole variety of ways, everything from turning back your thermostat to driving at a lower speed, or putting insulation in your attic. Energy conservation efforts can include what we call energy efficiency, but is not necessarily limited to that.

Energy efficiency, more specifically, is the technological way in which we use energy, that is, how much useful work we get out of a given energy input. It is really a matter of technologies and the systems that we employ. If you look at the substantial improvements in energy use over the past 20 years, you find that most of it is accountable from energy efficiency improvements; long term changes in technologies such as making our buildings more weather tight, putting in more efficient light bulbs, and driving more efficient cars. Energy efficiency is more important than specific conservation measures.

Global Enery Strategies: Living with Restricted Greenhouse Gas Emissions, Edited by J.C. White, Plenum Press, New York, 1994

Many of the kinds of things that were popular back in the 1970s, such as driving at 55 miles per hour or wearing a sweater when you are at home in the evening, are a lot less popular today, given the fact that energy prices are lower.

However, technological improvements in energy efficiency have continued to advance, albeit at a somewhat slower pace, since energy prices declined. It is important to get those concepts straight and to understand that, just as efforts to invest in energy supply technologies are an important feature of economic competitiveness, so are efforts to invest in the most energy efficient equipment that is available.

The following papers are going to deal with the three key sectors in which energy is used in our economy, transportation, industry, and buildings. We will get a sense, I think, from each of the three presenters of what enormous progress has already been made, as well as the technological potential that is possible in the future.

I want to make sure that we don't neglect other aspects of the energy use equation because, while end-use energy efficiency is certainly the most important element of energy efficiency, there are also important efficiencies in the way in which we convert energy from unusable forms to more useful forms and in the way in which we transmit and distribute energy.

For example, in power generation technologies, very substantial improvements are underway right now. Where we had conventional steam-based, coal burning generating plants stuck at an efficiency of about 35 percent, we are now in the midst of a revolution; moving to advanced combined-cycle plants, generally run on natural gas, which are already at efficiencies of 45 percent and appear to be headed towards 50 to 55 percent or more by the end of this decade.

The technological potentials of the kinds of efficiencies that can be improved in the transmission, distribution, and generation systems begin to build on each other. You keep building efficiency throughout the system and end up with energy systems that are radically improved in terms of their overall efficiency and certainly radically improved in terms of their emissions.

Take one simple comparison. Today you may be lighting your home with conventional incandescent bulbs. The electricity from those bulbs may be coming from a conventional coal-fired power plant which emits a given amount of carbon dioxide per lamp. By switching to a more efficient compact fluorescent light bulb, which uses about one-quarter as much energy to produce the same amount of light, and also switching to a more efficient combined cycle power plant, you receive the same amount of light, while producing 90 percent less carbon dioxide. There is a cumulative effect you can achieve by adding all of these things together.

THE TRANSPORTATION SECTOR

Steve Plotkin

U.S. Congress
Office of Technology Assessment

I am going to focus on personal travel in this paper, primarily because transportation is so complicated that I really can't hope to cover it all in this very short amount of time. But freight is clearly just as crucial an area of transportation, especially because trucks provide most of the freight transportation in the world.

In the United States, we have a very balanced freight system: trucks account for approximately 20 percent. But if you go to Western Europe, trucks account for over 60 percent of the freight transportation energy. The reasons for the difference are many, but the primary one is that the United States ships very large quantities of bulk products over great distances.

It is very important to recognize that, in not including freight, I am leaving out a very big chunk of the transportation market.

There are two very important perspectives on transportation.

> First: transportation is the source of social and environmental ills -- pollution, loss of ecosystems, green house emissions, loss of life and limb and noise pollution.
>
> Second: transportation is a key to economic progress and social, cultural and recreational opportunity.

Both are important. In the Third World and developing nations, they don't really care very much about the first perspective. They care about the second most of all.

Sometimes in the United States, some of the presentations I have heard about transportation seem primarily to focus on the first point, which is that transportation seems to be the source of all evil. Clearly, both perspectives are correct and both must be taken very carefully into account in designing transportation policy.

Before I get into the meat of this paper, I would like to state some of my conclusions.

First, I think we have got to understand that most people in this world do not travel very much and it is not because they don't want to. They do not have mobility, but as economies develop, the amount of travel is going to grow exponentially. That means that unless we do something very strong about transportation energy use, greenhouse emissions are going to grow enormously.

The second conclusion is that, although there are global things we can do about transportation and there are certainly technologies that can be applied globally, most solutions

Global Enery Strategies: Living with Restricted Greenhouse Gas Emissions, Edited by J.C. White, Plenum Press, New York, 1994

for transportation are very local and they are very personal: local in the sense that they are very, very dependent on local conditions. Very broad-brushed schemes for transportation energy solutions are really not going to work very well at the individual level. You have got to be very careful and examine each case in turn.

Another conclusion is that there is a bias against the automobile, probably a fairly justifiable one, in most of the conservation community. And there are an awful lot of people in that community who want to get us off of the automobile as fast as possible.

The problem is that the automobile is seen throughout the developing world as a status symbol and as a symbol of a civilized nation and of a growing and vital economy. Around the world, you see that the first thing people buy when they start to get rich is a car. And for us to go around preaching that the automobile is bad and you should try to go to mass transit systems and avoid automobiles probably isn't going to work.

It may be justified. It may be the really right thing to do, but it probably won't work and we are going to have to be very careful in the message we try to give to developing nations.

But this leads to a policy dilemma. It implies that perhaps one of the things we ought to do is to try to design a vehicle that works very well in the Third World and is very efficient, but that is sort of a dual-edged sword, because if we design a vehicle which works extremely well, the availability of that vehicle may push automobile development even faster than it is going without it. I am not really sure what to do about that, but it is certainly a problem worth considering.

Also, a look at Third World nations and their cities and how incredibly congested they are leads me to believe that electric cars, coupled with a renewables-based electricity system, is a solution that is worth a very careful look. We probably need a better battery, but the kinds of driving that are done in these cities, the amount of miles you can go in several hours is so low that an electric car, even with today's batteries, may work very well.

The critical issue here may not be the energy density of the batteries, it is just making sure that batteries last long enough and are inexpensive enough to be practical.

The next to last conclusion is that public transport certainly deserves a fair but a very cautious hearing; cautious because our experience in this country with mass transit has been pretty dismal and even though the transit systems in, say, Western Europe, work very well, in reality, there is a certain amount of fleeing from public transport in Western Europe. Over the last decade or two, Western Europe's automobile share has grown very rapidly and total travel is growing very rapidly as well. So, it is not clear that mass transportation is going to be the answer.

Finally, I think it is very important for us to realize that a lot of transportation policies just don't go with each other. Consider a policy to reduce congestion and, therefore, have

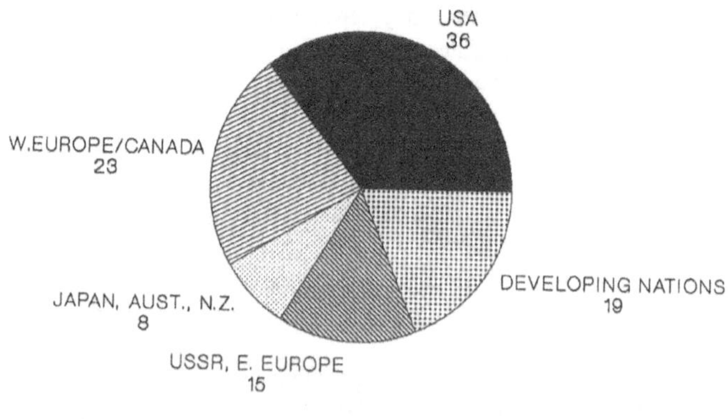

FIGURE 1. TRANSPORTATION CARBON EMISSIONS SHARES BY REGION, PERCENT

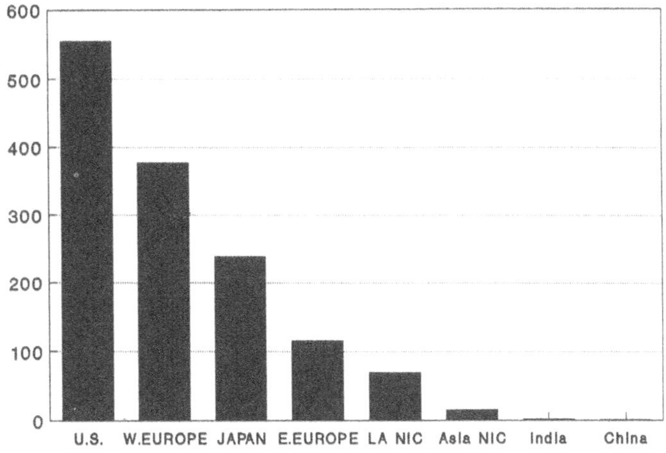

FIGURE 2. DIFFERENCES IN PASSENGER VEHICLE DENSITY CARS/1000 PEOPLE

free-flowing traffic and avoid quite a lot of energy waste from congestion. That kind of a policy cannot work with a policy that seeks to push public transportation, because public transportation really cannot compete with a free-flowing automobile system. Sometimes we are going to have to pick and choose.

Figure 1 shows how transportation carbon emissions are shared throughout the world. The United States, not surprisingly, certainly has the lion's share. U.S. travel is growing, although quite a bit slower than elsewhere in the world. There have been several projections that vehicle miles traveled (VMT), in this country will slow considerably from what our historical level has been.

Over the past several decades, the VMT growth level has been about 3 percent per year, but the most recent projections from the Energy Information Administration and other sources, are less than 2 percent per year growth in travel.

The only problem is that so far the expected slowdown is not happening, U.S. travel growth over time has been remarkably stable, showing no sign of slowing down.

In Western Europe, travel is increasing more rapidly than in the United States, and it's becoming increasingly auto-dominated, with suburbanization taking hold. In other words, Western European travel patterns are moving closer to U.S. travel patterns. And the rest of the world, especially the poor nations, pose this question: what will happen here when economic development really takes hold?

U.S. transportation emissions account for about one-third of all U.S. fossil fuel CO_2 emissions. For world transportation emissions, we are talking about 20 to 25 percent of fossil fuel CO2. It is not enough, of course, to worry only about CO2 because there are several other very important greenhouse gases, such as methane and chlorofluorocarbons, so that fossil fuel CO2 only accounts for about half of the total greenhouse problem in the long term.

In the conclusions, I talked about the future growth of transportation greenhouse emissions. Here is the problem. Comparing the U.S. and Western Europe to the rest of the world, the one thing you realize very quickly is that most people do not own cars, (Figure 2) but that gap probably will soon begin to be filled as Third World economies grow.

China's contribution doesn't even appear in Figure 2. It's vehicle use is virtually non-existent, at least their automobile is virtually non-existent, but if one examines the regions in China that are developing rapidly, the entrepreneurs who are making the money are buying automobiles. In Eastern Europe and China, in the Latin American developing nations, especially in the Asian developing nations, the rate of car ownership (Figure 3) is growing extremely rapidly.

Some projections are that by 2010, U.S. transportation CO2 emissions will be up about 25 percent, but in the developing world, they will be double or more. In fact, by 2010, the greenhouse emissions of the developing world could match the greenhouse emissions of the United States, because of the very substantial difference in growth rates of travel and in automobile ownership.

I would like to briefly review some of the methods we have to reduce greenhouse emissions, but I would like to point out that in earlier papers, here, there was some talk of a no-regrets policy. The other day I heard a talk, somebody mentioned the no-regrets policy and defined it as follows. He said the no-regrets policy is that policy about which the particular interest group advocating it has no regrets.

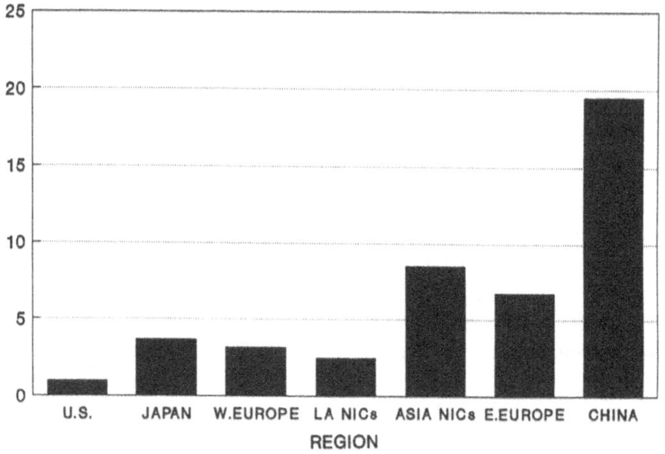

FIGURE 3. GROWTH RATES OF VEHICLE OWNERSHIP
PERCENT/YR GROWTH, 1975-87

I think it is perhaps a little naive to assume that the reason we are not doing some of the things that people want us to do is that we just haven't been thinking about the problem or we have been asleep at the wheel. All of the strategies involved in saving energy, especially saving transportation energy, involve certain kinds of tradeoffs. It may not be monetary. It may be simply a tradeoff between something like fuel efficiency and the ability to go zero to sixty a little faster.

From the conservation community's standpoint, perhaps, going zero to sixty faster than eight seconds seems a little silly and probably is, but the people buying automobiles want to do it. So, we have got to design strategies that will either convince people to forego some of these amenities or we are going to have to try to design around them.

But, anyway, getting back to these different options, first, the obvious one is to raise vehicle technical efficiency in automobiles and trucks. Second, changing vehicle fuels, is where we might be able to gain some efficiency from the fuels being considered. Methanol and natural gas have fairly high octane levels and other characteristics that let us design engines with a higher thermodynamic efficiency than an engine that uses gasoline. Besides that, some of these fuels may have fuel cycle emissions of greenhouse gases that are inherently lower than gasoline emissions would be.

Third, shifting to more efficient modes. Shifting to mass transit, bicycles, or even walking, and shifting from trucking to rail. We have to be very careful about some of these shifts, however. For example, it is fairly common to say that rail transport for freight is much, much more efficient than trucks, but the level of efficiency gained from a shift from truck to rail is going to be far less than is usually advertised. The reason is that the energy comparison is between all of the freight shipped by truck and all of the freight shipped by rail. The reality is that these are very different kinds of freight. Generally on rail you ship bulk commodities that are heavy and dense. On trucks you ship high value commodities that often must be delivered quickly. Often they are not dense and they inherently are difficult to ship with high energy efficiency. If you ship them on rail, the average efficiency on rail would be a lot lower than what the total average efficiency of all freight shipped on rail is today.

The fourth option is simply to reduce travel and that gets back to the two perspectives I talked about before. One of the things you can do is to reduce the need for travel; that is, you can let people enjoy the same services that they enjoy today with less travel requirement. And that focuses on the perspective about travel being a means to a better life. The other way you can reduce travel is simply by making it more difficult to travel, making it more expensive or putting restrictions on it. That reduces mobility and it does clean up the environment, but it also hurts people's lives and a lot of people will object very strenuously. We can improve traffic conditions because we lose a lot of oil with people sitting in traffic. That is also a dual-edged sword because when we improve traffic flow, often what we do is we increase total travel. So, we have got to be pretty careful with this option. And, finally, we can raise load factors with special lanes and other things like that.

Let me first talk briefly about high efficiency vehicles. One of the things to recognize in trying to improve the efficiency of our fleet is that technology isn't everything. Consumer choice and fleet turnover are extremely important. If we are constantly improving the fuel efficiency of the fleet, we want to do it in such a way that the vehicles are attractive, so people buy them and get rid of their old vehicles quickly. When we think about changing the new vehicle fleet, we want to be very careful to make sure that we change it in ways that don't turn people off, because we are just shooting ourselves in the foot if we do that.

In attempting to improve fuel economy, there are tradeoffs with performance and power, with cost, with weight, and with amenities such as power equipment, four-wheel drive, et cetera. If you load up the vehicle with things like four-wheel drive or power windows and door locks or other things, you reduce fuel efficiency; with higher performance, you also reduce fuel efficiency. We really may be asking people to make these tradeoffs if we want to get maximum fuel economy out of our fleet.

There is a tremendous controversy about what we can accomplish technology-wise with the current automobile fleet-- just how high can we go, say, by the year 2001. We have a range of estimates, which start with the auto companies' estimate that you can probably reach 30 or 31 miles per gallon by the year 2000 and all the way up to, the ACEEE, American Council for an Energy Efficient Economy, which is talking about a little over 40 miles per gallon by the year 2000. In our study, OTA felt that a reasonable level of fuel economy to reach by 2000 was probably in the range of 35 or 36mpg, but that is an arguable point.

Probably just as important as what the target for fuel efficiency should be is how do you regulate it? The current standard, which is 27.5 miles per gallon, actually works very poorly because it penalizes the automakers who make large cars and lets automakers who make primarily small cars get off virtually scot free. If, in fact, you examine the track of automobile fuel economy for the individual companies over time, you see that the American companies have been forced by the fuel economy standards to raise their fuel economy every single year. Some of the import companies have been able to actually lower their fuel economy over time because they started with primarily small cars.

You can substantially improve the fuel economy of the fleet without downsizing. A vehicle was introduced in 1992, which gets 44 percent more fuel economy than the vehicle it replaced and, yet, basically it is the same size (Figure 4). It has got the same power. It feels the same to drive, for the most part. It is a subcompact, and probably not a perfect model for the fleet. It has a standard transmission and a lot of the subcompacts sold today have automatic transmissions. It actually accelerates a little faster than the average subcompact, and is a little smaller inside.

- '92 Civic is larger, slightly faster
- '92's Efficiency technologies:
  - Weight reduction (5 percent)
  - Improved aero
  - Low resistance tires
  - "Smart" alternator
  - Better oils
  - VTEC engine
- Comparison of fuel economy:
  - California: 54.0 v. 40.8 mpg UP 34 PERCENT
  - 48 State : 59.0 v. 40.8 mpg UP 44 PERCENT
- Only California version will meet '94 US NOx std.

based on preliminary Honda data

FIGURE 4. 1992 HONDA CIVIC VTEC-E COMPARISON WITH 1991 CIVIC

I adjusted the miles per gallon of this car to see what would happen if it more closely resembled how the average subcompact performed (that is, I compensated for the difference in size, transmission type, and acceleration) and the 59 miles per gallon went down to about 50. That is an interesting number because the National Academy of Sciences had a value for their subcompact potential for 2001 of 39 miles per gallon and here is a vehicle which as a model for the subcompact fleet <u>today</u>, achieves around 50 miles per gallon.

Another way to deal with greenhouse gases is to go to alternative fuels. Again, they can give you higher efficiency and they can also reduce greenhouse gases just from the fuel cycle standpoint. They may have lower carbon content and they may emit smaller amounts of other greenhouse gases.

One of the major problems we face in getting alternative fuels into the market, however, is that gasoline has got almost a hundred years of consumer acceptance behind it. It is entrenched in the market and it also is just about the cheapest processed liquid you can buy in this country.

So, to get alternative fuels into the market, the market is not going to do it by itself. We need a government push of some kind, either regulatory or economic incentive.

Unfortunately, although alternative fuels have been praised very vigorously by various advocates, most of their benefits, although they are achievable, are far from automatic. For example, it is claimed that we would have a substantial energy security benefit from some of these fuels. Many of these fuels would have to be imported or, even if they don't <u>have</u> to be imported, under the most likely scenario, they would be. For example, methanol is more likely to come from Saudi Arabia than it is to come from this country.

The emissions benefits are far from clear. Although it is possible to generate reasonably low emissions with these fuels, many of the vehicles that have been tested do not generate particularly low emissions, especially with high mileage.

Finally, we have a law on the books in this country that says that if an auto manufacturer produces alternative fuel vehicles, he gets a CAFE (Corporate Average Fuel Economy) credit. That means that he can build a less efficient fleet than he would otherwise have done if he builds large numbers of these alternative fuel vehicles. In other words, you could end up losing in reduced efficiency what you gain in replacing oil use with the alternative fuels, with little net reduction in overall oil consumption. So, we have to take a very careful look before we continue to push these vehicles into the market.

Finally, we have to be careful in introducing these vehicles that we do not "poison the well"; by that I mean we do not introduce vehicles which are not fully ready for the market. There have been technologies, such as the heat pump, which were introduced many decades ago and then had to be withdrawn from the market because they were introduced without being ready for the market either technologically or from a service standpoint.

We can't do that with alternative fuel vehicles.

Using Public transportation is another way to save fuel. But when we examine what the U.S. has accomplished with public transportation, it is very difficult to make a case that we have saved much fuel.

In fact, basically, our public transportation system has been a failure. And by that, I mean, we have not been able over the past few decades to increase market share. Although we have tried greatly to increase ridership, we have been fairly unsuccessful doing that. Also the embodied energy in some of the more elaborate systems, especially some of the newer rapid rail systems, is so great that it isn't really clear that we are saving energy at all.

There are some real questions in looking at public transportation as it interacts with urban structure. Public transportation works best, and certainly rapid rail transportation works best, in extremely dense cities. One of the nice things about dense cities is that there is so much opportunity in terms of recreational, cultural, and job opportunities close at hand, that the amount of travel is much lower than in an auto oriented city like Dallas or Houston, for example.

Many people in the conservation community hope that by introducing rapid rail transit into cities, we will encourage them to become more dense and, therefore, most of the energy savings will not come from just shifting one automobile trip to one transit trip. What will happen is that as the cities become more dense, there will be a multiplication factor; that is, perhaps for every new transit trip we have, we will reduce automobile trips by a very high factor, perhaps as much as seven or eight. However, we have little proof that introducing a public transportation system of this nature will actually accomplish this goal.

There are many different policies that we can pursue. One potential policy that is now getting a great deal of attention is what we call full cost pricing; it involves recognizing that the use of gasoline has many externalities (social costs) associated with it that are not included in its price; that is, the accidents, the congestion, and the air pollution. If we throw these into all the transportation systems, we will tend to encourage systems which have fewer of these externalities and, hopefully, fewer greenhouse gas emissions.

Obviously, we probably will see another attempt in this Congress to try to get new fuel economy standards pushed through, a clear option for improving efficiency.

Urban planning is a critical option for efficiency if it is combined with transit systems, but it is not really clear that introducing transit alone will help the city grow more dense. However, there may be a possibility for introducing planning and transit in conjunction with each other; that is, very careful urban planning, designed to density a city, coupled with good rapid rail systems, that might work to greatly improve the efficiency of transportation and to reduce the need for travel.

THE POTENTIAL FOR REDUCING CARBON EMISSIONS THROUGH IMPROVED EFFICIENCY IN INDUSTRIAL PROCESSES

Marc Ross

Physics Department
University of Michigan

OVERVIEW

The potential for reducing carbon emissions through efficiency improvement in industrial processes is large in terms of technical capabilities. Simple cost-effectiveness considerations suggest that much can and should be accomplished even with present technology at present energy prices, and that much more should be accomplished with moderate energy, or carbon, taxes. Moreover, with new products and increasing affluence, the composition of production is changing such that industrial energy and materials consumption are growing more slowly than the economy. In addition, technological progress tends to reduce the overall energy and materials requirements for each industrial process. This relatively rosy scenario faces serious difficulties, however: industrial decision makers strongly discount future expenditures for energy for a variety of reasons. In addition, many slow-growing heavy industries may not have the capital and technical capabilities to invest in the best new production processes. For these reasons it is important to develop public policies to encourage the development and dissemination of more-efficient process technologies, and to assist energy-intensive industries to modernize, for example through utility demand-side management progams.

THE SHIFTING COMPOSITION OF PRODUCTION

The structure of industrial production is changing, and since energy intensities vary widely among sectors this is important.

The manufacture of bulk materials is much more energy intensive than manufacturing in general, and the use of materials in tons per year is declining relative to total production in all affluent societies (Williams et al. 1987).

The consumption of materials used in large quantities - paper, fertilizers, synthetic fibers, plastics, other industrial chemicals, cement, glass, pottery, and metals - is declining

Global Enery Strategies: Living with Restricted Greenhouse Gas Emissions, Edited by J.C. White, Plenum Press, New York, 1994

relative to total industrial production. This phenomenon, called dematerialization (Herman et al. 1989), has contributed a decline of 0.5-1.0% per year in the aggregate energy intensity of industry in industrialized countries.

The only bulk materials whose consumption still grows as fast or faster than the economy in the United States are in the chemical family: plastics and industrial gases. But growth rates for these materials as for all the bulk-chemical groups, have been falling dramatically. Consider, for example, what is happening to plastics production, saturation effects apply even to plastics.

Markets for heavy consumer products are saturating. For example, while the application of plastics to motor vehicles is increasing, unit sales of vehicles are no longer growing with the economy.

Innovative consumer products tend to have a low materials-to-cost ratio (kilograms per dollar). For example, although electronic equipment typically has a plastic structure or body, the ratio of weight-to-cost is low.

Materials are being used more efficiently. For example, linear low-density polyethylene, introduced in the late 1970s, allows the use of thinner films and is taking over low-density polyethylene markets.

This increasing efficiency is being driven, in part, by the competition among materials. For example, the competition for materials for grocery bags and beverage containers is fierce, putting a premium on efficient design and even beginning to bring in considerations of plastics recycling.

Improved materials are increasing product durability. Plastic pipes, other uses of plastic in construction, and plastic auto parts which reduce rust and corrosion often contribute to longer product life.

THE FAST-GROWING HIGH-TECHNOLOGY SECTORS

Another structural development is taking place among some of the low-energy intensity sectors. Statistical information and case histories support the concept that high-technology product sectors will play an important role in reducing the aggregate energy intensity of industry in the long term (Ross and Fisher 1992). During the 1970s and 1980s, the fastest growing sectors, especially electronic equipment, drugs, and instruments, were the main sources of growth in overall industrial production in the U.S. It seems likely that these and other research-intensive, innovative sectors will continue to propel growth in the U.S. economy as it matures.

The sectors with real gross output growing roughly 4% per year and faster for 1971-1985 are furniture and fixtures (excluding household furniture), printing and publishing, drugs and toiletries, rubber and miscellaneous plastics products, computer and office equipment, miscellaneous industrial machinery, electronic equipment, miscellaneous electrical equipment, selected military equipment, and instruments.

The key energy-related behavior is the low and declining energy intensity of these high-growth sectors. (Here, energy intensity can be measured as energy per unit of value added in a base year. Relative changes in energy intensity are, however, much better measured in terms of the ratio of energy to deflated gross output.) The decline in the energy intensity of these sectors is extremely rapid. This decline is not due to energy conservation efforts. It is due to continuing product innovation embodying design improvements which have a major impact on the value of the product but relatively little impact on the materials and energy requirements. It is also due to the rapid creation of new, more-modern, production facilities.

THE ENERGY, OR CARBON, INTENSITIES OF INDIVIDUAL PROCESSES

In principle, the energy intensities of most industrial production processes can be reduced to zero or near zero. Thermodynamically, the difference between a collection of materials and the same materials now shaped and assembled is nil. Only endothermic chemical processes require substantial energy use. The most important endothermic reactions are reduction of metals ores -- i.e., removing the oxygen - certain organic reformations, and dissocation of brines, such as separating the chlorine from salt, a modest part of all industry.

In practice, energy intensities are substantial even where no energy is needed in principle. But, historically, established energy-intensive processes, like smelting aluminum or making cement, have shown substantial ongoing reductions in energy intensity, even in the period 1958-1970) when energy prices were low and falling. Energy intensities of course fell more rapidly in the period 1971-1985.

There are, then, two sources of improved energy efficiency in making a particluar product: autonomous change, not associated with energy prices, and energy conservation associated with the price of energy.

AUTONOMOUS EFFICIENCY IMPROVEMENT

Fundamental change in production processes offers many major opportunities for improving energy efficiency and reducing greenhouse gas emissions. Some examples of these opportunities are listed in Table 1. Some are developments underway; others are still subjects for research and development. Having said that there are major potential benefits for energy use and carbon emissions, an important qualification is needed for policy making. For almost all manufacturing, the selection of the production process is not sensitive to energy price. For example, steel firms do not choose electric arc furnace technology with a primary focus on electricity prices, but because the scrap-based process involves much lower capital costs than the ore-based process, scrap is available in the region, the scale of production can be kept small, product markets look promising, etc. Empirical study supports this conclusion: electricity price has not had a substantial effect on this process choice, although it has influenced the particular arc-furnace parameters (Karlson and Boyd 1992). The general conclusion is even stronger for less-energy-intensive sectors than steel.

Table 1. Examples of potential fundamental process change with important energy and CO_2 consequences. Chemicals, paper, and food processes

Improved separation processes based on membranes, adsorbing surfaces, critical solvents, freeze concentrations, etc.
Ethylene chemistry based on natural gas feedstocks
Waste reduction using closed systems
New and improved catalysts for chemical processing
Recycling paper and plastics (i.e., into new material and product areas)
Continued improvements in the processing and forming of materials

Metals processes

Recycling post-consumer scrap (i.e., into new product areas)
Near net shape casting
Surface treatment with electromagnetic beams
Direct and continuous steelmaking
Coal-based aluminum smelting

General manufacturing

Oxygen instead of air processes/oxygen separation
Sequestering CO_2
Sensors and controls
New motors and controls
Low-energy welding
New adhesives
Alternatives to organic solvents for cleaning

Substantial evidence shows that general progress in production technology has benefits in all aspects of production. The energy costs associated with established types of processes are reduced along with other costs (Solow 1957, Berg 1978). There is the evidence just mentioned from energy-intensive industries in thc 1950s and 1960s when energy prices, including electricity prices, were low and, often, falling. New applications of electricity continue to occur, however, with surface treatment and specific heating (as contrasted with general volume heating in a typical oven) two currently important areas (Schmidt 1984). In addition, some electricity consumption is associated with added environmental controls. Thus, most current electrification is based on markedly superior production technology rather than price effects. Learning or experience curve studies of particular processes show fairly steady and remarkably large cost reductions over long periods (Joyce 1991, Argote and Epple 1990).

Energy Conservation as Such

Projects at an industrial facility which are largely motivated by energy-cost reduction tend to be small or medium sized (up to a few tens of million dollars) and to not involve fundamental process change. Let us examine this kind of project in a little detail because it is the most amenable to accelerated implementation.

The conservation supply curve, or CSC, is designed to exhibit the potential for such energy savings. The fundamental concept of a CSC is illustrated in Figure 1. The y-axis, K/S, is the unit capital cost of a project, the cost divided by the rate of energy saving (e.g., $ per average kW, or $ per bbl/day, saved). The X-axis is the cumulative energy savings capacity, either an absolute rate (e.g., average kW, or kWh/yr) or a percentage.

Alternatively, analysts often convert the project's costs into a stream of annual payments divided by the annual energy savings, so that the y-axis becomes energy price (Meier 1983). An unpublished estimate of the electricity conservation opportunity in fabrication and assembly sectors is shown in this form (Figure 2) for purposes of illustration. The solid curve represents typical investment behavior, with the financial criterion represented by a capital recovery factor (CRF) of 33% per year, which effectively characterizes conservation investment by typical manufactuers. According to this CSC, there is a relatively small opportunity for further conservation at the present price of about 5¢/kWh. That is, the electricity intensity is close to its equilibrium value, given the present price and investment behavior. However, with a criterion more representative of society's time horizon (CRR = 16%), a 30-40% reduction of electricity intensity would be justified at the present price (dashed curve) and achieved, after some delay.

Fuel Switching

In addition to energy conservation, or efficiency improvement, as such, carbon emissions can be reduced by fuel switching. At industrial sites the potential is limited

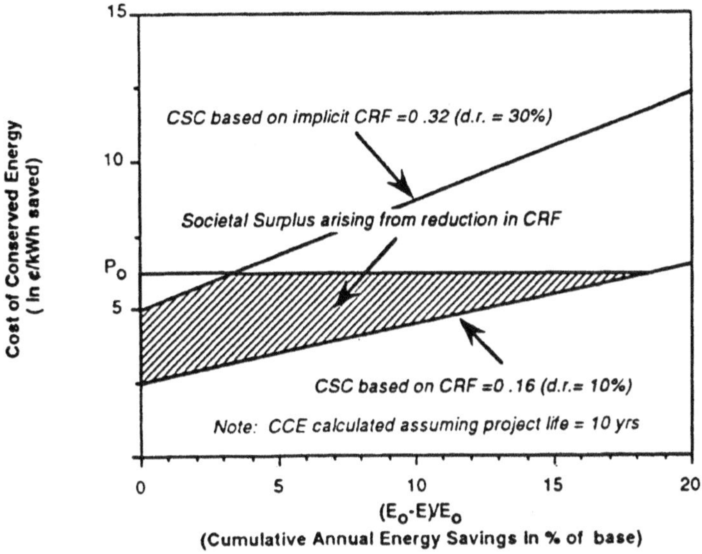

FIGURE 2. CONSERVATION SUPPLY CURVE IN TERMS OF ENERGY PRICE WITH DEPENDENCE ON CAPITAL RECOVERY FACTOR

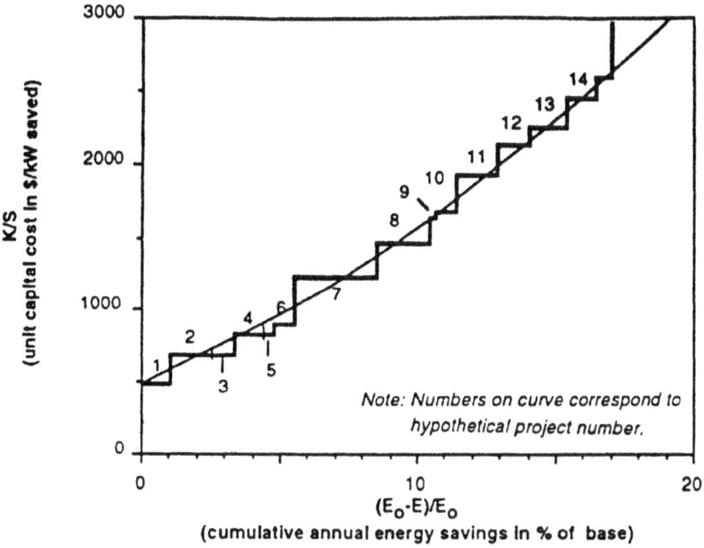

FIGURE 1. FUNDAMENTAL CONSERVATION SUPPLY CURVE

because the lowest carbon fuel, methane, is already dominant in the United States. Much could be accomplished in other countries by switching to this fuel, which also often has the lowest overall cost.

Coal and petroleum consumption by industry is also substantial, but these fuels are often used where substitution with natural gas is difficult. The main use of coal in the U.S. is in iron and steel, where substitution would be a major technological challenge. We know how to reduce iron ore with coal, but not with other fuels. Petroleum consumption is large in three areas: at petroleum refineries as an aspect of refining, as a feedstock to make plastics and other organic products, and in remote locations.

Electricity use by industry is heavy, so shifting electricity generation away from coal would be helpful, as would making generation and delivery more efficient. Switching from direct use of fuel at the factory to use of electricity may thus be useful in the long run. In the forseeable future, only applications where electricity use is much more efficient are attractive for reducing carbon emissions. Substition of natural gas for coal-generated electricity is also attractive, where the application is efficient.

INHIBITIONS TO THESE FAVORABLE CHANGES AND POLICIES TO EXPEDITE THE CHANGES

Inhibitions

Developing countries are still in the stage of rapidly growing use of bulk materials, with their associated energy- and carbon-intensive industrial processes. Will these countries be able to embrace modern and efficient process technologies as their prodction expands, or will they be led into the old paths? In a study of China's steel industry, I found this a complex issue (Ross 1991). Not only are old processes used but old product lines are being bought by customers. China has an enormous iron market, with cast iron products prominent in agriculture and households. This is inefficient compared to use of steel in most cases. Moreover, steel products are often of low quality in China, so that thick sheet needs to be used in making trucks, for example; and important opportunities are inihibited, like the building of high-pressure pipelines for natural gas, in part because the quality of steel piping is inadequate.

Energy- and carbon-intensive industries in industrialized countries are mature. The use of their products has peaked. This good news was discussed above, but it's also bad news. The bad news is that technical expertise and the ability to invest in new processes in these industries is declining. For example, many of the professors in iron and steel metallurgy have recently retired; and they are not being replaced. Just at a time when new process technology is needed, capabilities for research, development, and innovative investment are disappearing.

As indicatcd above, energy-conservation projects in industry are subject to strong discounting of future benefits. Small or even moderate cost savings are not that critical to marketing strategy in spite of what is taught in undergraduate economics. Most industrial activity is in perpetual flux. Management must act to create new products, improve product capability, meet new production schedules, develop effective approaches to marketing, improve the effectiveness of the work force, meet new regulatory requirements, and perhaps enter new areas of business. Mere cost savings can get priority only if: 1) their potential is large, or 2) they can be achieved in conjunction with more important goals.

Policies

More support is needed for research and development on new processes (and products) in the mature energy-intensive industries. In the United States, the federal government has been almost the exclusive patron of research in pure science, in military areas, and to a large extent in medicine, agriculture, and electricity supply. The government has also helped fund the development of certain high-technology products such as supercomputers. But it leaves to the private sector most of the research and development in areas related to the core activities of society, where technology has great impact on the environment - such as materials production. The private sector cannot create and implement the new technology needed by society in these mature areas. Changed governmental priorities are called for (Ross and Socolow 1991).

Policies to create a balance between investment in energy supply and conservation are needed. The reduction of energy-supply subsidies in needed. The involvement of the energy industry, such as electric and gas utilities, in investment to improve their customers' efficiency will help. This can bring the utilities' long-term financial criteria into play.

Energy or carbon-emissions taxes will help as part of a package of policies. Increasing the prices for energy is a policy with some desireable features, but it is a weak spur for investment, as discussed above. It can be a supplement to other policies focused on investment. Judicious use of tax revenues to assist firms in making process-efficiency investments might be very effective.

Market incentives and regulations to encourage products manufactured with low greenhouse gas implications needs to be explored as a policy area. Regulations requiring accurate labelling of recycle content in products, and government procurement standards for such products, are examples.

Many potentially useful policy options were tried in the late 1970s and early '80s; but unfortunately, they have not been evaluated. How well such policies work often depend on their precise formulation and implementation. Nevertheless, there is reason for optimism. We have learned in the past two decades that public policy is an effective and appropriate way to deal with energy demand. Effective government initiatives to encourage energy supply are well known. Just as appropriate a policy target is energy consumption. We need to learn more about exactly which public policies are most appropriate in the many different contexts of demand.

REFERENCES

Argote, L. and D. Epple, 1990. "Learning Curves in Manufacturing," Science 247, 920-924.

Berg, C. (1978) Science 199, 608.

Herman, R., Siamak, A.A. and Ausubel, J.H. (1989("Dematerialization," Technology and Environment, eds. Ausubel, J.H. and Sladovich, H.E. (National Academy Press, Washington).

Joyce, W. (1991) "Energy Consumption Spirals Downward in the Polyolefins Industry," Energy and the Environment in the 21st Century, ed., Testor, J.W. (MIT Press, Cambridge, MA), pp. 427-435.

Karlson, S. (1990) The Impact of Energy Prices on Technology Choice in the U.S. Steel Industry (Department of Economics, Northern Illinois University, DeKalb, IL), working paper.

Meier, A., Wright, J. and Rosenfeld, A.H. (1983) Supplying Energy through Greater Efficiency (Univ. of California Press, Berkeley).

Ross, Marc and Ronald Fisher, 1992. "Will Per-Capita Industrial Energy Consumption Grow in the long Term," unpublished working paper.

Ross, Marc and Feng Liu, 1991, "The Energy Efficiency of the Steel Industry of China," Energy 16, 833-848.

Ross, Marc and Robert Socolow, 1991. "Fulfilling the Promise of Environmental Technology," Issues in Science & Technology, Spring, 61-66.

Solow, R.M., 1957. "Technical Change and the Aggregate Production Function," The Review of Economics & Statistics 39, 312-320.

Williams, R., Larson, E. & Ross, M. (1987). "Materials, Affluence and Industrial Energy Use," Ann. Rev. Energy 12, 99-144.

INCREASING ECONOMIC GROWTH AND REDUCING CARBON EMISSIONS THROUGH IMPROVED ENERGY EFFICIENCY

Arthur H. Rosenfeld

Center for Building Science
Lawrence Berkeley Laboratory

INTRODUCTION

Recent studies highlight the large potential for electricity and fuel savings through improved energy efficiency. These studies show that E/GNP (primary energy use/$ of GNP) could be reduced by nearly 50% for the U.S. for optimum economic growth over a period of 10-20 years, the typical turnover time for automobile, appliance, and equipment stocks and the time needed for building retrofits. This paper discusses recent progress and potential future savings in automobiles, refrigerators, windows, lighting, and through urban heat island mitigation. These energy savings are translated into savings of carbon dioxide (CO_2).

BACKGROUND

Total energy consumption in the U.S. has fluctuated dramatically during the last 30 years. These fluctuations are divided into three recent energy eras in **Figure 1:** pre-oil embargo (1960-1973), high oil prices (1973-1986), and frozen efficiency (1986-1989). Between 1960 and 1973, prior to the oil embargo, energy use and U.S. gross national product (GNP) were inexorably linked and climbed at a rate of about 4% per year. However, during the 13 years of high oil prices following the 1973 oil embargo, national energy use stayed constant, while U.S. GNP grew by a total of 35%, or 2.5% per year. When oil prices collapsed in 1986, gains in energy efficiency improvements slowed or stopped and energy consumption began climbing again at a rate of about 3% per year. In 1990, consumption was virtually unchanged from 1989, due primarily to mild weather, slow economic growth, and higher energy prices following the Iraqi invasion of Kuwait in August (U.S. DOE, 1991).

Significant energy savings have occurred since 1973. By 1990, the difference between GNP-projected and actual primary energy use was about 33 EJ, a reduction in energy intensity (energy/GNP) of 38%. Structural change toward a more service-oriented economy is credited with one-fourth to one-third of this change, resulting in energy efficiency-related savings between 22 EJ and 25 EJ (Schipper et al., 1990 and U.S. OTA, 1990). However, increasing concerns over national economic competitiveness and global environmental degradation (Rosenfeld and Price, 1992) underscore the need for further improvements in energy efficiency.

RECENT PROGRESS AND FUTURE ENERGY SAVINGS

Figure 1 showed that overall U.S. energy consumption relative to gross national product (E/GNP) decreased 35% between 1973 and 1985. However, this aggregate value obscures whether this improvement is due to structural change or to energy efficiency and conceals both sector-specific variations and individual technological advances that have occurred. In fact, the efficiency of many individual energy-using products typically doubled during this time. This doubling occurred with minimal investments; the price difference between the existing and the more efficient product was typically recouped through reduced energy bills in 3 years or less. There are many examples of this efficiency doubling effect.

Automobiles

Since the Corporate Average Fuel Economy Act (CAFE) took effect in 1975, new car fuel economy has doubled from 14 miles per gallon (mpg) to 28 mpg. At a fleet average of 14 mpg, the 150 million U.S. automobiles, light trucks, and taxis on the road in 1990 (Davis and Hu, 1991) would have required 9 million barrels of oil per day (Mbod). Today, the fleet average has reached 21 mpg. When full turnover occurs and the fleet average is equal to the new car fuel economy of 28 mpg, these same 150 million vehicles will require only 4.5 Mbod[1]. The total savings of 4.5 Mbod is equivalent to the sum of the pre-war production from Kuwait (1.8 Mbod) plus Iraq (2.8 Mbod), or about 3 times the production of Alaska's Prudhoe Bay, or 15 times the expected production of the Arctic National Wildlife Refuge (ANWR).

Figure 2 shows that improving fuel efficiency from 14 mpg to 27-28 mpg costs about $300 (retail), but saves about 400 gallons/year, worth about $500 (at $1.25/gallon), so the simple payback time is under 1 year (Difiglio et al., 1990; Ledbetter and Ross, 1990). Alternatively, investments in oil supply, such as the extraction of oil at Prudhoe Bay, have simple payback times closer to ten years.

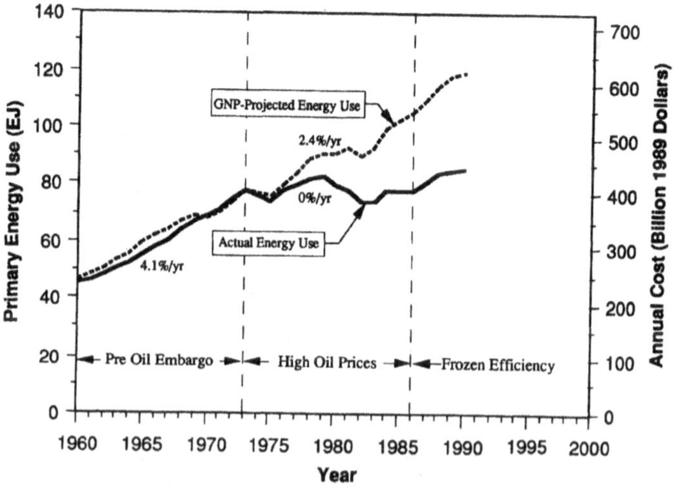

FIGURE 1. U.S. ENERGY CONSUMPTION

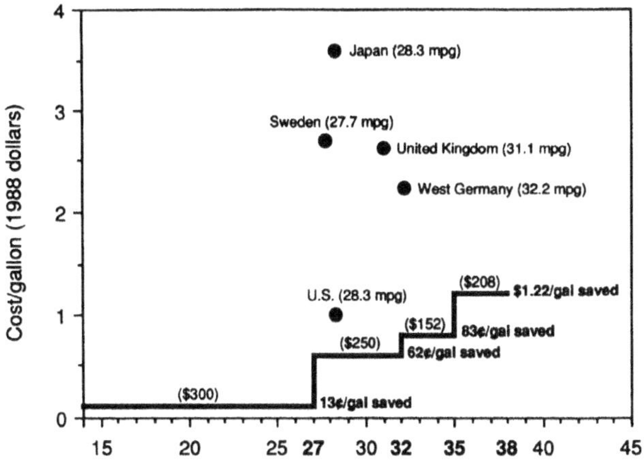

FIGURE 2. NEW CAR FUEL ECONOMY (MPG)

REFRIGERATORS

Between 1972 and 1985 the number of refrigerators sold in the U.S. climbed 33% and their average volume grew, but their efficiency doubled and their energy use decreased as a result of first California and then national efficiency standards. In 1972, U.S. refrigerators using an average of 1726 kWh/year required electricity from 40 large power plants (1 GW each, selling 5 BkWh/year). Efficiency improvements have reduced the electricity required for today's average refrigerator to 930 kWh/year, provided by only 20 large power plants. Annual avoided CO_2 emissions from this energy savings are 70 million tonnes (Mt).

Figure 3 shows that today's 930 kWh/year refrigerator, which retails for about $100 more than its 1726 kWh/year ancestor, saves this additional cost through reduced energy bills in 2.5 years. Savings realized after the first 2.5 years are then available for other investments, adding to economic growth. In contrast, the 20 power plants avoided through refrigerator efficiency improvements would have tied up their investor's money (and the nation's economic growth) for more like 25 years.

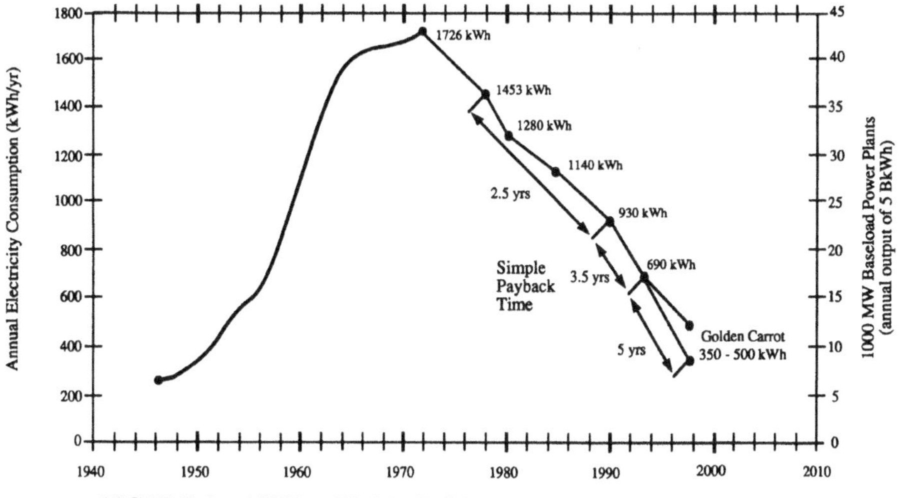

FIGURE 3. AVERAGE ENERGY USED BY REFRIGERATORS

More efficient refrigerators are currently under development in response to a "Golden Carrot" incentive program directed at U.S. refrigerator-freezer manufacturers. This program uses financial incentives to encourage development and production of products that exceed existing efficiency standards. A consortium of utilities, government agencies, refrigerator manufacturers, and other interested parties has pooled $30 million to be paid to the winning manufacturer. The consortium's goal is to encourage commercialization of refrigerator-freezers that beat the 1993 National Appliance Standards by between 25 and 50%, approaching 350 to 500 kWh/year for an 18-cubic-foot unit. The units must also be CFC-free, and extra points are given for designs without even HCFCs (L'Ecuyer, 1991; ACEEE, 1991; Wald, 1992). When these efficient units have saturated the market, another 10 large power plants, emitting 35 Mt of CO_2 per year, will not be needed.

WINDOWS

In 1973 more than half of the residential windows in the U.S. were still single-glazed (i.e. had a thermal resistance of R-1 [sq.ft.-hr.-°F/Btu]) and less than half were double-glazed (R-2). High prices for heating fuels following the oil embargo triggered the development of low-emissivity (low-E or "heat-mirror") gas-filled double- or triple-glazed windows, where R-4 is now standard, and R-8 is being introduced in colder areas. **Figure 4** shows that 1990 sales of 20 million low-E windows (10 ft^2 each) represents about 30% of the residential market and 15% of the commercial market. Industry sources predict that sales of these windows will grow to between 55 and 80% of the residential market and between 30 and 50% of the commercial market by 2000.

Each low-E window costs about $20 more retail ($10 more wholesale) than double-glazed windows, but saves 10 million Btu (MBtu) over its lifetime of 20 years. When market saturation of these efficient windows has been reached, the savings will be equivalent to the production of 30 offshore oil platforms (10,000 bod, with a capital cost of about $500 million each). Annual avoided CO_2 emissions from these energy savings will be 18 Mt.

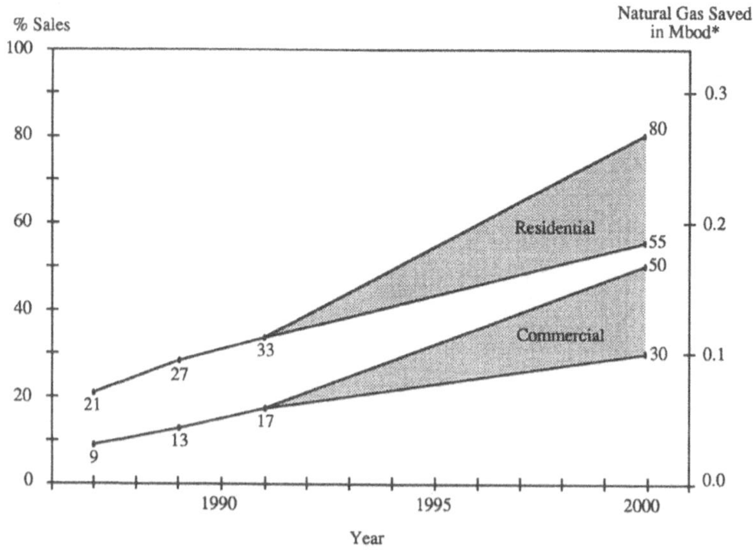

FIGURE 4. SAVINGS FROM USE OF LOW-E WINDOWS

LIGHTING

Another technological advance that was triggered by the oil embargo was the development of electronic ballasts that then ushered in compact fluorescent lamps (CFLs). Together, these new technologies will save about 200 BkWh annually when market saturation is reached, equivalent to about 40 large power plants emitting 140 Mt of CO_2 per year.

In the U.S., one 16-watt CFL replaces a series of about twelve 60-watt incandescents because it burns twelve times longer. This CFL, which costs about $10 more retail ($5 more wholesale), will save 440 kWh over its 40-month life in a commercial building, and has a simple payback time of less than 1 year. **Figure 5** shows actual and projected CFL sales in the U.S. Recent experience has shown higher lamp sales in areas where utilities, under the provisions of new profit rules associated with integrated resource planning (IRP) and demand-side management (DSM), have actively promoted their sales through rebate programs.

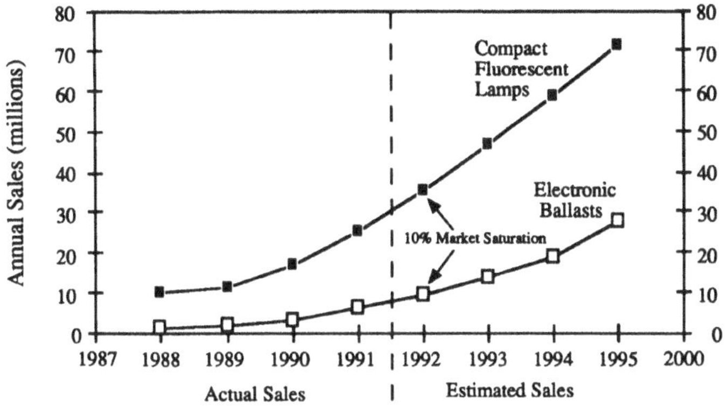

FIGURE 5. SALES OF COMPACT FLOURESCENT LAMPS

IRP is a novel approach to utility energy resource planning that puts supply-side and demand-side resources on a level playing field (Krause and Eto, 1988). Utilities then make decisions on how to provide required energy services (warmth, illumination, motive power, etc.) by combining array of options including construction of new power plants, renovation of existing power plants, cogeneration, purchases of power from independent power producers, and promotion of energy efficiency. About 35 states throughout the U.S. have some sort of IRP regulations (Levine et al., 1991).

DSM programs aimed at reducing customer demand for energy services through improved energy efficiency have developed as a direct result of IRP. To promote DSM programs, regulatory commissions have often removed existing utility financial disincentives. Traditionally, utilities reaped little benefit from promoting energy-efficient technologies; indeed, such technologies lowered energy sales. Now, utilities in 12 states have revised profit-making rules to permit recovery of some DSM program costs and utilities in 10 states have positive incentives for DSM activities. It is estimated that U.S. utilities spent close to $2 billion on DSM programs in 1990, with some utilities investing between 2 and 6 percent of their operating revenues in these programs (Edison Electric Institute, 1990).

Following adoption of DSM incentive mechanisms, the New England Electric System now projects that 40% of its future demand can be met through DSM programs. Two California utilities (Pacific Gas and Electric Company and Southern California Edison) expect to meet about 75% of their needs during the next decade through DSM (Levine et al., 1992). DSM programs have also recently been adopted in Canada, Thailand (International Institute for Energy Conservation, 1992), Brazil (Geller, 1991), and Western Europe (Mills, 1991). A number of other developing and Eastern European countries have become interested in such programs.

URBAN HEAT ISLAND MITIGATION

"Urban heat island" is a term describing the tendency for cities to become hotter than their immediate surroundings due to the presence of dark surfaces and small amounts of vegetation. Since about 1940, U.S. cities have been growing hotter every summer. These urban heat islands are the result of replacing vegetation with dark-colored streets, parking lots, and roofs. The Los Angeles basin is a particularly troublesome heat island because the

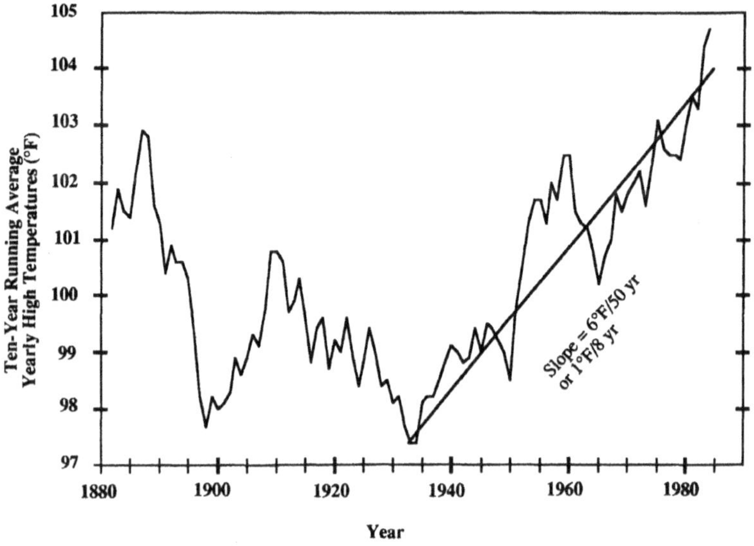

FIGURE 6. YEARLY HIGH TEMPERATURES IN LOS ANGELES

extra 7°F there makes the smog significantly worse. Between 1880 and 1940, Los Angeles actually cooled about 2°F as water and orchards were introduced to the area. The cooling gains were reversed when trees were cut and dark-colored asphalt roads, parking lots, and roofs were constructed. **Figure 6** shows the historical dynamics and the currently growing heat island in Los Angeles. Since 1940, summer average temperatures in Los Angeles are up about 7°F and are continuing to rise by about 1°F per decade. The 7°F rise since 1940 requires Los Angeles basin utilities to supply an additional 1000 MW of peak power for air conditioning.

Research indicates that at extremely low cost we can cool cities significantly, save energy, and reduce pollution through planting urban trees, vines, and using lighter-colored surfaces (Akbari et al., 1988; Akbari et al., 1989). Trees improve the urban climate and, through shading and evapotranspiration, reduce summer cooling energy use in buildings at only about 1 to 10% of the capital cost of the avoided power plants and air conditioning equipment. Using light-colored roof paints and lightening asphalt with white aggregate and a layer of sand are even more effective. **Figure 7** shows large measured temperature differences of about 70°F between white and black surfaces and of about 40°F between concrete and asphalt. This figure also shows that surface temperatures of a hypothetical light-roofed "green city" are expected to be 15°F less than an average city and that temperatures of a hypothetical "white city" where there is less urban vegetation (such as in dry cities like Phoenix) are predicted to be another 10°F less (Taha, et al., 1992; Bretz and Rosenfeld, 1992).

In addition to reducing electric bills, urban vegetation and white surfaces slow the growth of atmospheric CO_2. Rural trees sequester carbon and avoid several tons of CO_2 per ton of wood, but one ton of shade trees reduces the need to burn 10 to 20 tons of coal or gas to meet air conditioning demand. Thus, urban trees are 20 to 50 times more efficient at offsetting CO_2 than rural trees because of the two-pronged benefits of sequestering CO_2 and reducing the need to burn fossil fuels for generating electricity.

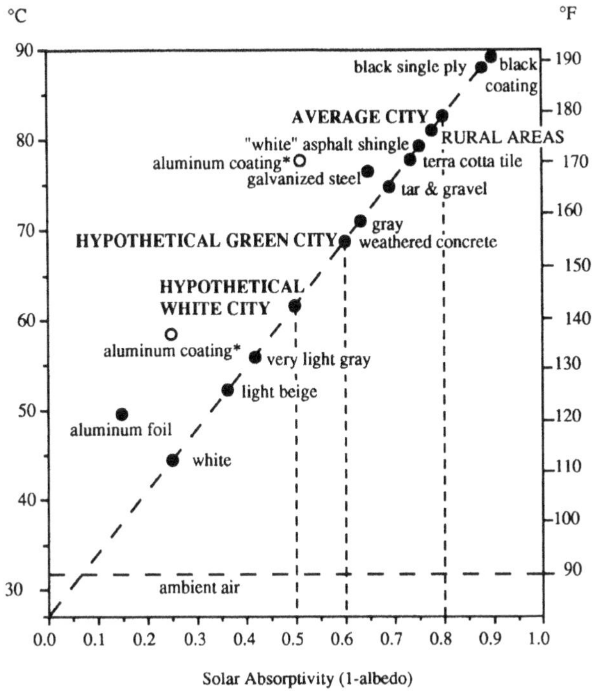

FIGURE 7. SURFACE TEMPERATURES OF DIFFERENT MATERIALS

TECHNICAL POTENTIAL FOR IMPROVED ENERGY EFFICIENCY

The conservation supply curve is a powerful analytical and planning tool for assessing the technical and economic potential of various energy-efficiency options. Conservation supply curves provide an economic ranking of available options by relating energy savings achieved by implementing a given option to that option's "cost of conserved energy" (CCE). On a conservation supply curve each efficiency option is plotted as a step in which the height is the CCE, the width is the annual energy saved, and the area under the step is the total annualized cost of investing in the efficiency option (Meier, et al., 1983).

Traditionally, conservation supply curves have been constructed to reflect one of two technology saturation levels: technical potential or achievable. The technical potential saturation level is based on engineering calculations without constraints on costs or assessment of the probability of successful implementation. Achievable saturation scenarios are based on options that cost no more than the cost of expanding energy supply and on actual experience with program participation rates (Krause and Eto, 1988). Even the most ambitious utility conservation programs have rarely captured more than 50% of the technical potential. However, with recently adopted profit incentive mechanisms, some utilities can now earn a percentage of avoided costs for conservation programs, so the level of technical potential that is actually achieved will undoubtedly be higher in the future.

Electricity Conservation Supply Curves

Analysts in the Energy Analysis Program at LBL have recently constructed a comprehensive electricity conservation supply curve for U.S. residential buildings (Koomey et al., 1991). This curve, which was derived using a extensive database of appliance efficiency and costs developed for the U.S. Department of Energy and a detailed analysis of thermal integrity measures in single-family dwellings, is shown in **Figure 8.** The supply curve evaluated the technical potential for electricity efficiency improvements using available or "advanced" technologies[2] and assumed a 7% real discount rate, and an analysis period of 1990 to 2010. For those measures costing less than the cost of electric power to residential customers, or 7.6¢/kWh in 1989, the technical potential for residential electricity savings in all buildings in 2010 is about 40%, or 404 BkWh of the projected 2010 residential frozen-efficiency baseline use of 1028 BkWh.

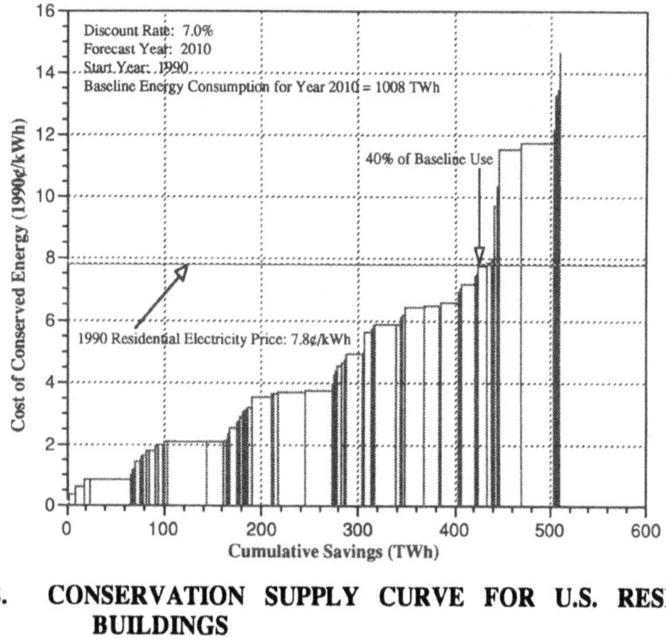

FIGURE 8. CONSERVATION SUPPLY CURVE FOR U.S. RESIDENTIAL BUILDINGS

Analysts in the Center for Building Science and the Energy Analysis Program at LBL also compiled nine potential conservation supply curves that depict the technical potential for electricity savings in U.S. residential and commercial buildings by about the year 2000 (Rosenfeld et al., 1992). Within this compilation, the 11-step EPRI conservation supply curve (Faruqui et al., 1990), with an additional first step for white surfaces and urban trees to save air conditioning, represented the approximate mid-range of the compiled supply curves. If all 12 options (identified in **Table 1** and **Figure 9**) were fully implemented in

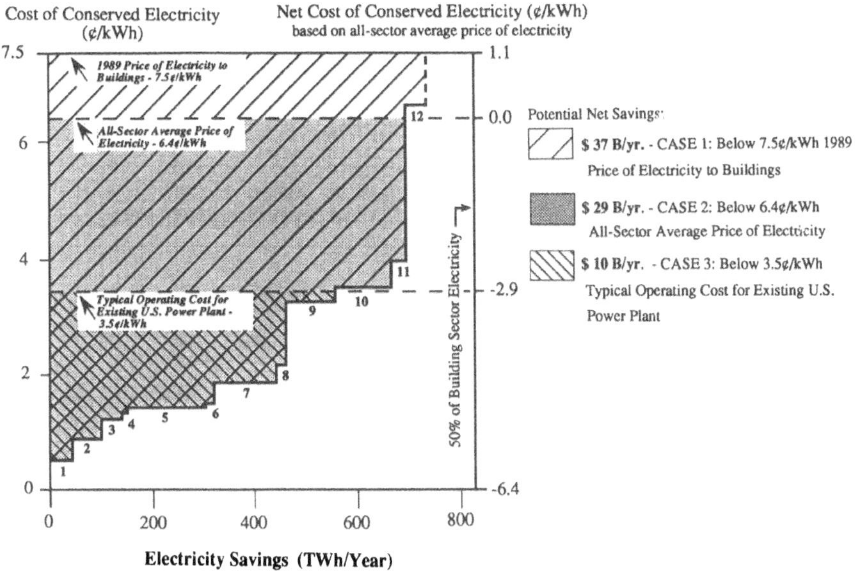

FIGURE 9. POTENTIAL SAVINGS FROM CONSERVATION

TABLE 1. THE COST OF SAVING ELECTRICITY AND CARBON DIOXIDE THROUGH CONSERVATION IN BUILDINGS

Measure	A CCE *¢/kWh* d=0.06	B Net CCE *¢/kWh* (A-6.4[a])	C Net CC CO_2 *$/tonne* (10xB/0.7[b])	D Potential U.S. Electricity Savings *(Cumulative TWh/yr)*	E Potential U.S. CO_2 Savings *(Cumulative Mt CO_2/yr)*
1 White Surfaces + Urban Trees[c]	0.5	-5.9	-84	45	32
2 Residential Lighting[d]	0.9	-5.5	-79	101	71
3 Residential Water Heating	1.3	-5.1	-74	139	97
4 Commercial Water Heating	1.4	-5.0	-72	149	104
5 Commercial Lighting	1.5	-5.0	-71	315	221
6 Commercial Cooking	1.5	-4.9	-70	322	225
7 Commercial Cooling	1.9	-4.5	-64	437	306
8 Commercial Refrigeration	2.2	-4.2	-60	459	321
9 Residential Appliances	3.3	-3.1	-44	562	393
10 Residential Space Heating	3.7	-2.8	-39	667	467
11 Com. & Ind. Space Heating	4.0	-2.4	-35	689	482
12 Commercial Ventilation	6.8	.4	6	734	514
• US Buildings Sector 1989 Use				1627	1140

a Represents value of primary energy imputs in production of electricity (11,500 Btu/kWh). Note: U.S. DOE's allocation of primary energy consumption between air conditioning and ventilation end-uses differs from EPRI's allocation on which savings in this table are based. EPRI's 1987 base case allocations are 1.77 for air conditioning and 0.89 for ventilation. When aggregated, the differences in the two base cases are minimal.

b Carbon emissions from Edmonds, J.A. 1989.

c Akbari, et al., 1988 and Akbari et al., 1989.

d EPRI measures 2-12 have been converted from 5% discount rate to 6% discount rate by multiplying EPRI CCE values by 1.05. EPRI values for "Potential U.S. Electricity Savings (Cumulative)" have been expanded by a factor of 1.4 to account for EPRI's failure to include improvements for which they had no cost data (residential residual appliances and commercial miscellaneous equipment) and to adjust the EPRI savings, which were compared to utility projections that included 9% naturally-occurring efficiency improvements, to frozen efficiency.

existing U.S. residential and commercial buildings, the cumulative savings would be 734 billion kWh which is 45% of 1989 building sector electricity use.

Savings from conservation supply curves can be evaluated through comparisons with specific electricity prices. All efficiency options that fall below the electricity price are profitable to society. Using the 1989 price of electricity to buildings of 7.5¢/kWh, all 12 options are cost effective and would result in an annual net savings of $37 billion. From a societal point of view, we can use the all-sector average price of electricity of 6.4¢/kWh as a cost-effectiveness threshold. From this perspective, the first 11 options are cost effective and would result in savings of 689 billion kWh at an annual net savings of $29 billion. Finally, using the operating cost for an existing U.S. power plant of 3.5¢/kWh, only the first 9 options are cost effective and savings drop to 562 billion kWh with an annual net savings of $10 billion.

Fuel Conservation Supply Curves

Unlike electricity, there are few conservation studies available for fuel use in buildings. Two recent studies of residential natural gas conservation potential indicate savings of about 50% are possible for residential natural gas use at less than the current average price of $5.63 per million Btu (Meier et al., 1983; SERI, 1981). Extrapolating this estimate to cover all gas and oil use in buildings yields savings of about 5.2 quads. Subtracting out the $10 billion annualized cost of the conservation investment results in a net savings of nearly $20 billion/year (Rosenfeld et al., 1992).

Carbon Dioxide Savings Potential

Electricity and fuel savings from conservation supply curves can be transformed into units of avoided CO_2 following conversions outlined in Rosenfeld et al., 1992. Using these conversions, potential CO_2 savings indicated by the LBL residential electricity conservation supply curve, which estimated electricity savings of 404 BkWh, are over 280 Mt. The EPRI supply curve, which estimated residential and commercial electricity savings of 734 BkWh, indicates the potential for 514 Mt of avoided CO_2. This is about 10% of U.S. 1989 emissions of 5 Gt CO_2. Potential carbon savings from fuel conservation are about 300 Mt of avoided CO_2, or about 6% of U.S. 1989 emissions.

CONCLUSION

These examples show that it has been easy to double the efficiency of a number of energy-using products and that significant energy savings are possible using today's most efficient technologies. Of course, the challenge is to make the market right for investments in these energy-efficient products. Market incentives must be tailored to overcome market failures. First, nothing will work without energy use labels. After that, standards should be developed to guide the market toward improving product efficiencies. Standards are also needed for areas where decision-makers don't consider investments in energy efficiency, such as businesses with relatively small energy costs or where payback periods are longer than 3 years. Incentives, such as the Golden Carrot program, should be coupled with standards to push efficiency even further. At the utility-level, IRP and DSM can have an enormous impact on the adoption of more efficient products, and regulatory commissions should encourage this type of planning for all utilities. Once these policies are in place and large numbers of energy-efficient products are being purchased, today's equipment will need to be replaced with more efficient models. To stay ahead of the market, we need to continue to finance energy efficiency research and development.

REFERENCES

Akbari, H., Huang, J., Martien, P., Rainer, L., Rosenfeld, A., and Taha, H., 1988. "The Impact of Summer Heat Islands on Cooling Energy Consumption and Global CO2 Concentration," *Proceedings of the 1988 ACEEE Summer Study on Energy Efficiency in Buildings* (5). Washington, DC: American Council for an Energy-Efficient Economy.

Akbari, H., Garbesi, K., and Martien, P., 1989. *Controlling Summer Heat Islands, Proceedings of a Workshop on Saving Energy and Reducing Atmospheric Pollution by Controlling Summer Heat Islands.* Berkeley, CA: Lawrence Berkeley Laboratory.

American Council for an Energy-Efficient Economy, 1991. *The Golden Carrot News.* Washington, DC: ACEEE.

Bretz, S. and Rosenfeld, A.H., 1992. *Mitigation of Urban Heat Islands: Materials and Utility Programs.* Presented at the *NIGEC Supercities Conference*, San Francisco, CA, October 28, 1992.

Davis, S. and Hu, P., 1991. *Transportation Energy Data Book: Edition 11* ORNL-6649. Oak Ridge, TN: Oak Ridge National Laboratory.

Difiglio, C., Duleep, K.G. and Greene, D.L., 1990. "Cost Effectiveness of Future Fuel Economy Improvements," *The Energy Journal*, Vol. 11, No. 1, January.

Edison Electric Institute, 1990. *State Regulatory Developments in Integrated Resource Planning*. Washington, DC: EEI.

Edmonds, J.A., 1989. *A Preliminary Analysis of U.S. CO_2 Emissions Reduction Potential from Energy Conservation and the Substitution of Natural Gas for Coal in the Period to 2010*, DOE/NBB-0085, February 1989.

Faruqui, A., Mauldin, M., Schick, S., Seiden, K., and Wikler, G., 1990. *Efficient Electricity Use: Estimates of Maximum Energy Savings*, (CU-6746). Palo Alto, CA: Electric Power Research Institute.

Geller, H.S., 1991. *Efficient Electricity Use: A Development Strategy for Brazil.* Washington, DC and Berkeley, CA: American Council for an Energy-Efficient Economy.

International Institute for Energy Conservation, 1992. *E-Notes: Quarterly Newsletter of the International Institute for Energy Conservation.* Washington, DC: IIEC.

Koomey, J.G., Atkinson, C., Meier, A., McMahon, J.E., Boghosian, S., Atkinson, B., Turiel, I., Levine, M.D., Nordman, B., and Chan, P., 1991. *The Potential for Electricity Efficiency Improvements in the U.S. Residential Sector,* (LBL-30477). Berkeley, CA: Lawrence Berkeley Laboratory.

Krause, F. and Eto, J., 1988. *Least-Cost Utility Planning: A Handbook for Public Utility Commissioners (v. 2): The Demand Side: Conceptual and Methodological Issues.* Washington, DC: National Association of Regulatory Utility Commissioners.

L'Ecuyer, M., 1991. Personal communication.

Ledbetter, M. and Ross, M., 1990. "A Supply Curve of Conserved Energy for Automobiles" *Proceedings of the 25th Intersociety Energy Conversion Engineering Conference*, Reno, NV, August 12-17, 1990, New York: American Institute of Chemical Engineers.

Levine, M.D., Gadgil, A., Meyers, S., Sathaye, J., Stafurik, J. and Wilbanks, T., 1991. *Energy Efficiency, Developing Nations, and Eastern Europe: A Report to the U.S. Working Group on Global Energy Efficiency.* Washington, DC: International Institute for Energy Conservation.

Levine, M.D., Geller, H., Koomey, J., Nadel, S., and Price, L., 1992. *Electricity End-Use Efficiency: Experience with Technologies, Markets, and Policies Throughout the*

World. Washington, D.C: American Council for an Energy-Efficient Economy. (also as LBL-31885).

Meier, A., Wright, J., and Rosenfeld, A., 1983. *Supplying Energy Through Greater Efficiency: The Potential for Conservation in California's Residential Sector.* Berkeley, CA: University of California Press.

Mills, E., 1991. "Evaluation of European Lighting Programs: Utilities Finance Energy Efficiency." *Energy Policy*; 19 (3): 266-278.

National Academy of Sciences, 1992. *Policy Implications of Greenhouse Warming: Mitigation, Adaptation, and the Science Base.* Washington, DC: National Academy Press.

Rosenfeld, A.H. and Price, L., 1992. *Incentives for Efficient Use of Energy: High Prices Worked Wonders from 1973 through 1985, What Are Today's Alternatives to High Prices?* To be published in the Proceedings of the POWER (Program on Workable Energy Regulation) Conference: *The Economics of Energy Conservation*, Berkeley, CA.

Rosenfeld, A.H., Atkinson, C., Koomey, J.G., Meier, A., Mowris, R.J., and Price, L., 1992. *A Compilation of Supply Curves of Conserved Energy for U.S. Buildings*, (LBL-31700). Presented at the *Western Economic Association International Conference*, San Francisco, CA, July 12, 1992. To appear in *Contemporary Policy Issues*, January 1993.

Schipper, L., Howarth, R.B., and Geller, H., 1990. "United States Energy Use from 1973 to 1987: The Impacts of Improved Efficiency." *Annual Review of Energy;* 15: 455-504.

Solar Energy Research Institute (SERI), 1981. *A New Prosperity: Building a Sustainable Future.* Andover, MA: Brickhouse Publishing.

Taha, H., Sailor, D., and Akbari, H., 1992. *High-Albedo Materials for Reducing Building Cooling Energy Use*, (LBL-31721). Berkeley, CA: Lawrence Berkeley Laboratory.

U.S. Department of Energy, 1991. *Annual Energy Review 1990*, (DOE/EIA-0384). Washington, DC: Energy Information Administration.

U.S. Office of Technology Assessment, 1990. *Energy Use and the U.S. Economy*, (OTA-BP-E-57). Washington, DC: U.S. Government Printing Office.

Wald, M.L., 1992. "Utilities Offer $30 Million for a Better Refrigerator". *New York Times*, July 8, 1992.

FOOTNOTES

[1] The calculation is for 1990's number and mix of automobiles, light trucks, and taxis (about 150 million) assuming that fuel economy for light trucks scales along with auto fuel economy.

[2] "Advanced technologies are those that have been developed and for which engineering estimates of performance are available. "Promising" technologies have not been included due to lack of data.

LONG TERM OPTIONS FOR ENERGY SUPPLY AND DEMAND SIDE MANAGEMENT

Tom Morron

Vice President, Customer Service & Marketing

and

Fred Denny

Vice President, Engineering Fossil Fuels
Edison Electric Institute

INTRODUCTION

A great deal has been said and written about future energy options and the need for responsibility and caution in protecting the world's natural environment. Clearly, energy policies and environmental policies are intractable connected and must be considered jointly.

This paper offers a vision of long-term options for energy supply and demand side management to meet world-wide energy needs in a manner that is technologically efficient, cost effective, and environmentally responsible. This paper bases many of its findings on recent studies or experiences reported in the United States.

The long-term, as defined here, extends from the present to the end of the next century. This paper suggests an incremental approach, beginning by identifying the best current technological options and advocating a series of possible transitions to arrive at a desirable future scenario. The thesis provided is one of systematic advancement rather than precipitous reaction or indecision and inaction given the complex and uncertain nature of the issues.

The supply side options addressed by this paper focus primarily on improvements in technologies for generating, storing, and delivering electricity. This paper indicates that achieving an efficient and effective energy future will require meeting an increasing portion of future energy needs through the use of energy in the form of electricity.

This paper addresses research, development, and demonstration efforts needed to efficiently use hydropower, coal, oil, and gas resources in the short-term. A future scenario for advanced nuclear power generation is described, and strategies are discussed for

Global Enery Strategies: Living with Restricted Greenhouse Gas Emissions, Edited by J.C. White, Plenum Press, New York, 1994

expanding the role of renewable or "alternative" generating technologies (solar energy, wind power, geothermal energy, fuelwood energy, etc.).

Energy storage possibilities are included in the context of more effectively employing the various generation technologies, and the benefits to be gained from generation and storage advances are compared with the benefits attainable from energy delivery advances.

The demand side options addressed by this paper presume continuing improvements in supply side technologies for generating, storing and delivering electricity. This presumption is singularly important in discussions which relate to successful achievement of the "technological leap-frog" in developing countries and rarely receives sufficient attention in international discussions. The paper will discuss this matter in some detail.

This paper addresses both supply options and demand side options because the advanced demand side electrotechnologies can provide significant environmental and economic benefits only if a sufficient supply of high quality power is available.

The paper will examine a number of the more advanced electrotechnologies currently available as well as a number which can be reasonably anticipated in the longer term which will produce environmental benefits net of additional fuel burn at the power plant. These examinations will focus on industrial, commercial, agricultural and residential applications and will feature market barrier assessments.

The basic examination will consist of niche market orientation in developing countries, isolation of one or more electrotechnologies appropriate to the task of the niche, a discussion of the performance characteristics inherent to the electrotechnology which would include waste management aspects, and a comparison to other alternatives in terms of fuel use and environmental consequences.

DEMAND SIDE MANAGEMENT

To many professionals in the energy industry, the term "demand side management" simply means the vigorous pursuit of energy efficiency. In the electric utility sector, it has also come to mean the pursuit of strategies to reduce, eliminate or even reverse growth trends in the usage of electricity. There is, however, a much broader meaning to the term which must be recognized as the nations of the world attempt to deal with global issues. that broader meaning encompasses the deployment of electric end-use technologies as replacements for existing fossil technologies for environmental purposes.

Just as demand-side management techniques can reduce or eliminate unnecessary use of energy and associated emissions, demand-side techniques can be employed which, while increasing energy use, can concomitantly reduce net environmental emissions. Targeted electrification is a demand-side management technique which must be recognized as a legitimate component of any overall demand-side management strategy. Failure to include such a component in a national energy strategy unnecessarily deprives that nation of not only a valuable tool with which to make a contribution to global environmental enhancement, but deprives it of an economically productive technology with which to raise the standard of living of its population.

Perhaps an examination of the energy experience of the United States would serve to illustrate how it is that conservation, energy efficiency and targeted electrification can peacefully coexist. Today, total United States emission of carbon dioxide (CO_2), are only 7% higher than in 1973. At the same time, the Gross National Product (GNP) of the United States has grown over 50% with nearly two-thirds of that growth fueled by a 380 million ton annual increase in coal consumption. Even though coal is generally categorized in the popular press as the "worst" global climate change offender, CO_2 emissions per dollar of Gross National Product in the United States have declined from four pounds per dollar in 1973 to just a bit more than two pounds per dollar in 1990. (1)

An analysis of the data very clearly indicates a strong association between increased energy efficiency and increased CO_2 efficiency or decreased global climate change impacts. The improved CO_2 efficiency of the United States' economy cannot be fully accounted for because of increased use of non-combustion fuels (specifically nuclear energy) and improvements in the fuel economy of cars. Improved fuel efficiency of the nation's auto fleet and increased use of nuclear energy (both reducing CO_2 emissions) account for only 11% and 12% respectively of the total improvement in CO_2 efficiency since 1973. (2)

The driving force behind this improvement has been a 54% increase in the use of electricity in the United States since 1973. (3)

The foregoing is not meant to suggest in any way the abandonment of demand side techniques which focus on reducing or minimizing electricity consumption. Quite the contrary. These demand side techniques contribute substantially to economic and environmental progress.

For example, in the United States, the equivalent of some 25,000 MW of generating capacity is expected to be displaced by the year 2000 through demand side management programs currently in place. (4) If one were to assume a conservative construction cost of approximately $1000/KW, this constitutes a foregone cost of some $25 billion. It would not be an extraordinary assumption to make in forecasting that all of these 25,000 MW would be nuclear and therefore, millions of toms of emissions would not be produced as a result of these programs. In fact, recent projections from the Electric Power Research Institute estimate CO_2 emissions reductions from these programs at approximately 76.5 million tons. (5) These economic and environmental benefits are well worth vigorous pursuit.

While the previously referenced conservation-type programs are reducing demand in the year 2000 for electricity and associated CO_2 emissions, it is estimated that targeted electrification may well increase electricity use by some 336,000 Gwh in the same period. What is dramatic is that the increase in electricity usage in the year 2000 will cause a decrease in CO_2 emissions of between 71 and 175 million tones net of the additional fuel burned at the power plants. (6) Thus, the combination of these two demand side techniques can produce not only billions of dollars in avoided costs, but reduce CO_2 emissions between 147.5 and 251.5 million tons as well.

What is being advocated here is a recognition that demand side management techniques can both increase and decrease electricity consumption in an economic and environmentally sound manner at the same time and that an energy policy planner must seek an appropriate balance between the two. To overstress one denies society the very substantial benefits inherent in the other.

ENERGY EFFICIENCY EQUALS ELECTRICITY

There are two kinds of energy efficiency: using energy more efficiently, and using more efficient energy. Each kind of efficiency has played a role in human progress, but of the two, using more efficient energy has produced the largest gains thus far.

Consider the history of illumination: the burning torch was followed by oil lamps and candles, then these yielded to a gas flame. Each step represents a more efficient use of the same kind of energy. It is the inherent efficiency of electricity that offers the greatest potential for our energy future. The reason is somewhat simple: electricity can substitute for less efficient forms of energy. In so doing, energy is saved even though more electricity is used.

It is a truism that societies should pursue activities directed towards maximizing overall energy efficiency. Disagreement occurs over which approach should be pursued. For some, the first target is to cut electricity use because of the energy losses which occur at the power plant when heat is converted into kilowatt-hours. Some have gone so far as to claim that the energy efficiency of electricity is analogous to "cutting butter with a chain saw." The myth that electricity is wasteful results from ignoring the efficiency with which electricity is actually used and the inefficiency with which fuels are used in the market place. (7)

For example, the best power plants convert about 40% of the energy consumed into electricity. (8) However, electric motors convert 90% of electricity into useful motion. (9) By comparison, even the most efficient automobile converts less than 20% of its fuel energy to a drive shaft. (10) In other words, the efficiency with which electricity can be used more than offsets the inefficiency of making electricity. Meanwhile, the efficiency with which fossil fuels can be used is remarkably low in most applications, and for reasons of fundamental physics, inherently limited. (11)

When primary energy forms are converted to electricity, the resulting kilowatt-hours can be applied to an almost infinite number of tasks. Thomas Edison once observed that "If the enormous energy latent in coal could be made to appear as electrical energy by means of a simple transforming apparatus, the mechanical methods of the entire world would be revolutionized." (12) Because electricity is the medium of the integrated circuit, electric-based activities are further enhanced directly by the ever-expanding power of the microprocessor. Computers and processors can easily monitor, record and control any process that used kilowatt-hours. Processes and devices that burn fuels require complex mechanisms to convert the information into electric impulses understandable to computers. For example, the specific manner in which gasoline is burned in the combustion chamber of a car engine is complex, exceptionally difficult to observe, and essentially impossible to monitor in a operating vehicle. The kilowatt-hours consumed in an electric motor can be precisely recorded, observed and controlled. (13)

From a policy perspective then, energy planners should strive to make the most efficient use of the most efficient energy, electricity. There are, however, a number of leading environmental groups in the United States and elsewhere who are calling for an absolute limit on the amount of electricity used in order to address concerns over possible global climate change. (14)

The siren song of simplicity has lead these groups to conclude that only those demand side techniques which reduce, eliminate or even reverse the growth of electricity use in the marketplace are appropriate for consideration. Growth in electricity use, they conclude, can only bring greater emissions from power plants to the detriment of the environment and depletion of non-renewable resources. No allowance for technological substitution or improvement is made in their recommendation.

This critical error in judgement has been made before. Recall the Club of Rome and their report issued in 1971 titled "The Limits to Growth." With the aid of electricity and computers, the study predicted that growth would be limited by the depletion of non-renewable resources and by the accumulation of pollution. Their suggested remedy was to stop growth immediately lest the world and humanity suffer dire consequences.

What has come to pass is a far cry from their predictions. According to their study, the world mined the last of its gold ore 11 years ago, the least of its mercury and silver seven years ago, and the least of its tin five years ago. Zinc ran out two years ago and the world will have exhausted its petroleum some time last year. Copper and lead will be exhausted this year.

The Club of Rome was not the first group or individual to make such predictions. The major "improvement" added by the authors of the study was the use of computers to make their predictions. As one of the study's critics commented dryly: "Malthus in, Malthus out."

When the authors of "Limits" predicted that the world would run out of copper this year, they did it by taking the rate of copper consumption 20 years ago and increasing that consumption by the expected rate of economic growth. Their model could not know that some 20 years hence, copper telephone wires would be replaced with glass fibers. These fibers are made from the most common element in the earth's crust and can carry thousands of times more information than the very best copper cable. The unaccounted for variable in their study was human ingenuity which, in the case of the world's copper resources, produced not only a technology which contributed to the preservation of a non-renewable resource, but a technology orders of magnitude more efficient than that which it replaced.

It is useful to note that this technology is made possible through an electrically powered process. Had the world followed the Club of Rome's admonition and stopped the growth of electricity usage along with everything else, this improvement and its attendant environmental benefits would have been denied us.

There are a vast number of new electrotechnologies whose performance characteristics are known to us today which, while increasing the consumption of electricity, can produce not only greater economy and productivity in commerce and industry, but produce environmental improvements devoutly desired by us all. Placing arbitrary limits on this most efficient and flexible energy forms works a hardship humanity can ill afford.

PRIMARY ENERGY REDUCTIONS THROUGH TARGETED ELECTRIFICATION

In order to understand the implications of using more efficient energy forms at the point of end use, the concept of a "primary energy use ratio" is most useful. This ratio facilitates the computation of fossil fuel energy savings due to targeted electrification. It is defined as the primary energy required to produce a unit of service or product through the use of fossil fuel at the end-use level divided by the primary energy required to produce the same unit of service or product through the use of electricity at the end-use level. (15)

Using the formulation, a primary energy use ratio of greater than one would indicate that electricity end use is more energy efficient than fossil fuel based end use. A ratio of two would mean that two times the primary energy is required to produce the same product or service if fossil fuel instead of electric end use is used.

The energy efficiency potential of electricity as a substitute for fossil fuels can be seen by calculating the primary energy use ratio for some representative end-use technologies which are familiar.

In the residential sector, a comparison of high efficiency electric heat pump (SEER = 16.4) with a high efficiency gas furnace (AFUE = 92%) yields a primary energy ratio of 1.27. This means that 27% more fossil energy is required to perform the same function as the electric application. In short, electricity is the more efficient energy form. (16)

In the commercial sector, a comparison of a moderate efficiency electric heat pump (COP = 2.6) with a moderate efficiency boiler (Eff. = 70%) yields a primary energy ratio of 1.27, again confirming electricity as the more efficient energy form. (17)

Again in the commercial sector, an examination of chillers reveals a COP range for absorption systems from 0.5 to 1.4 and COPs for electric-input chillers at above the 6.0 level. The comparison yields a primary energy ratio of approximately 1.5. (18)

In the industrial sector, producing a ton of steel through electric arc melting produces a primary energy ratio of 2.7 when compared to the use of coal for the same purpose. Primary energy is conserved even though the use of electric energy increases. (19)

For illustrative purposes, the following is the calculation of the primary energy ratio for the use of an electric plasma cupola in the melting of steel.

To produce one ton of ferrous castings in an electric plasma cupola, it takes 200 KWh of electricity and .084 tons of coke. Total primary energy, then, equals 200 KWh x 100,000 Btu per KWh + (.084 tons of coke x 2000 lb. per ton x 13,000 Btu per lb.) equals 4.18 million Btu.

To produce one ton of ferrous castings in a conventional cupola, it takes 0.22 tons of coke. Total primary energy, then, equals 0.22 tons of coke x 200 lbs. per ton x 13,000 Btu per pound equals 5.72 million Btu.

The primary energy ratio, then, is found by dividing 5.72 million Btu by 4.18 million Btu. The resulting ratio is 1.37. (20) Again, this ratio simply means that primary energy is conserved even though the use of electric energy increases.

In the transportation sector, the comparison of an electric bus to a diesel powered bus yields a primary energy ratio of 5.63. In this case, the diesel bus requires 771 Btu per passenger mile while its electric counterpart requires only 137 Btu per passenger mile. (21)

In fleet use, the comparison of a gasoline powered, Chrysler mini-van with the same vehicle powered electrically produces a primary energy ratio of 1.74. The gasoline vehicle requires some 9400 Btu per mile while the electric vehicle requires only 5400 Btu per mile. (22)

In passenger vehicles, a comparison of a gasoline powered car with an electrically powered car produces a primary energy ratio of 2.36. the gasoline vehicle requires some 6524 Btu per mile while the equivalent electric vehicle requires just 2764 Btu per mile. (23)

These examples are not meant to imply that all end use applications of electricity result in primary energy conservation. They are portrayed here to illustrate the necessity of targeting electric end use applications. When properly applied, fossil energy can be conserved by using the more efficient energy form, electricity.

USING ELECTRICITY TO REDUCE CO_2 EMISSIONS

Electric steel making offers one of the clearest and most tantalizing examples of using kilowatt-hours to cut both energy usage and CO_2 emissions. The electric melting process deposits essentially all of the energy directly in the melt where it is needed with virtually no energy wasted at the point of use. The energy lost in making the electricity is substantially less than the energy wasted by a typical blast furnace. (24) The difference between the two processes is analogous to using a fire to boil water as opposed to placing electrodes directly into a mug of water.

The United States produces over 200 billion pounds of steel a year. For every pound of steel made electrically, roughly 1.3 pounds of CO_2 are eliminated. This accounting considers the elimination of the coal burned and CO_2 emitted in the fuel cycle for the blast furnace, and assumes that only coal is burned to make the necessary electricity. (24) This calculation is based on a CO_2 emission rate of 1.9 x 10-4 lbs. CO_2 per Btu of coal burned. (25) Similar calculations have been made using a CO_2 emission rate of 2.2 x 10-4 lbs. CO_2 per Btu of coal burned with a result showing 2.2 pounds of CO_2 eliminated for every pound of steel made electrically. (26)

While no attempt was made to account for the actual market penetration of this electrotechnology in the U.S. steel industry, the magnitude of the CO_2 reductions possible through the use of electricity should be of more than passing interest to the energy planner.

Using ultraviolet light in place of a natural gas fired oven to dry paint offers significant energy benefits as well as the potential for significant CO_2 emission reductions. Calculations done by Science Concepts, Inc. comparing the energy and CO_2 characteristics of using natural gas ovens vs. ultraviolet light to dry and cure the paint on a new car show a 92% energy reduction and a CO_2 reduction of about 2 lbs. per car. (27) Given just the U.S. annual new car production in the millions of vehicles, the potential for both energy and CO_2 savings in just this one application is substantial. The motivation, however, for using

UV paint drying has not been energy or emission savings. Rather, UV dryers are up to ten times faster and produce a more uniform and thus higher quality finish. (28)

Ultraviolet light can also be used to dry the ink on paper. Here again while energy and CO_2 savings are significant, the motivation for its use are productivity and economic benefits. Using the calculations from Science Concepts which compare UV drying with gas fired ovens in the drying of 4000 pages of magazine print, energy savings of some 62% result along with between 1.4 and 1.7 pounds of CO_2 reductions result. (29) Given literally billions of magazine pages printed around the world, the magnitude of these savings cannot be understated. UV drying can also be used for a range of activities not only in the printing industry, but in the electronics industry for protective coatings on circuit boards, in the wood particle board industry, and in the hardening of coatings on no-wax floors. (30)

Glass can be produced electrically in a fashion similar to the processes used for metals. Glass making can be entirely electrified or by means of a process known as electric boosting, partially electrified. Electric glass making provides improved product quality, smaller facilities and lower environmental impacts than conventional fossil fired technologies. (31) It is estimated that this process offers a 69% energy saving along with between 0.7 pounds and 1.8 pounds of CO_2 reductions for every 12 standard bottles produced when compared with a conventional gas fired oven. (32)

Electric vehicles also offer significant CO_2 reduction potential. In the United States, the average car is driven approximately 10,000 miles per yea. If that car were powered electrically, CO_2 emissions would be reduced by some 3666 lbs. when compared to a gasoline powered car. (33) With approximately 180 million automobiles in use in the United States, even a small penetration of this electrotechnology would produce significant environmental benefits.

It is worth noting that in each of these examples, the emission reductions shown are net of the additional fuel burned to produce the electricity required as well as transmission and distribution losses in delivery to the point of use.

There are, of course, many other electrotechnologies which offer not only primary energy savings, but environmental advantages as well. Each must be examined for its particular market niche as no one technology is the panacea for the world's environmental or energy problems. It is clear, however, that concentrating demand-side efforts on only those techniques which reduce, eliminate or reverse growth in electricity usage is an inappropriate dna unbalanced approach to dealing with our common desires for energy efficiency and environmental improvement.

GENERATION DECISIONS

Generation mix decisions must be made on the basis of many factors including capital costs, fuel and maintenance costs, environmental protection, current mix available for base load, cycling, and peaking, customer load shapes, customer power quality needs, etc. The term "integrated resource planning" is sometimes defined to include many of these factors. Increasingly, power from independent power producers and other competitive trends must also be taken into account. More than ever before committing to unneeded new capacity can prove to be very costly, and having supply fall short of demand can be fatal.

When a sufficient amount of electric generation capacity is available, that is when an adequate generating capacity margin is available, opportunities exist to optimize operating economics, maximize system reliability, and minimize environmental impacts. These objectives can be accomplished by making intelligent decisions about unit commitment, maintenance scheduling, and system dispatching.

COAL POWERED GENERATION

Coal-fired electric generating plants currently provide about 55 percent of the electricity used in the United States. Moreover, taking economic and practical considerations into account, it is inescapable that coal must continue to be a significant part of the generating mix. Coal resources are abundant, and an extensive infrastructure currently exists to support its use as a boiler fuel. This infrastructure includes and depends on coal miners, rail and other modes of transportation, and of course, power plant facilities.

The opportunities to increase the efficiency of using coal fired generation and reduce environmental impacts fall into two categories:

1. Opportunities to increase the efficiency of conventional, pulverized coal units, and
2. Opportunities to use the new "clean coal technologies."

Improved generation efficiency and innovative power generation technology can greatly reduce pollutant formation. This is particularly important in the developing countries. China, for example, is undoubtedly going to rely primarily on coal resources to meet future energy demands and to fuel economic growth. The unresolved issue is how will China and other countries with major coal resources use their resources.

Relatively small improvements in generating unit efficiencies or heat rates can increase energy production and reduce pollutant levels per unit of coal burned. The generating units in some of the Eastern European countries currently have significantly poorer heat rates than the units in more modern industrial nations, and straight forward engineering efforts, using current technology can improve efficiencies. With improvements in metallurgy and increased use of computers and control equipment even more efficient steam cycles and heat rates can be achieved.

Clean coal technologies are projected to achieve air quality compliance at lower capital, operating, and maintenance costs than other current technologies. While scrubbers reduce only SO_2 emissions, some clean coal technologies offer simultaneous reduction of both SO_2 and NOX emissions. In addition, given the need to comply with SO_2 limitations, some clean coal technologies produce lower CO_2 emissions than conventional, pulverized coal technologies.

To date, private industry in the United States has spent more than $1 billion to develop clean coal technologies. If on-going clean coal technology demonstration projects are successfully completed a variety of clean coal technologies should be available for commercial order by the late 1990's.

GAS POWERED GENERATIONS

In the United States, gas-turbines comprise approximately 8 percent of operable capacity for electric generation, and about 9.5 percent of total net electricity generation is provided by gas. Looking to the future, it is likely that natural gas will increasingly be used as a fuel to produce electricity. Natural gas powered generating facilities offer several advantages including small incremental additions, relatively low capital costs, short lead time construction, and relatively low short-term financial risk. From the point of view of environmental protection, gas is also seen by many utilities as a preferred option.

The increased use of natural gas, in both combustion turbines or in combined cycle systems, will undoubtedly play a major part in meeting future U.S. and world-wide electricity needs. Current technology allows gas turbines to reach efficiencies of approximately 35 percent. Gas turbines with secondary steam cycles operating as a "combined cycle" can achieve efficiencies as high as 45 percent. Further on the horizon, new technologies such as the "humid air cycle" promise still higher efficiencies, simpler operation and increased reliability.

A great deal of effort is currently being directed toward gas regulation, transportation, and contracting issues. New technologies are being studied to unlock unconventional gas reserves.

OIL POWERED GENERATIONS

In the United States, oil currently is used as the primary energy source to produce approximately 4 percent of total net electricity generation.

Oil continues to be vulnerable to supply disruptions and there is a great deal of uncertainty about future oil prices. Most U.S. utility companies do not plan to have oil play a major part in future generation mix scenarios.

NUCLEAR POWER

In the United States, nuclear power facilities account for approximately 14.6 percent of operable capacity for electric generation. The continued and increased use of nuclear power can make a significant contribution to meeting future energy needs with minimal adverse environmental consequences. To accomplish this electric utilities must continue to operate plants with utmost attention to safety and reliability, and government must work toward predictability and stability in the licensing process and assure that a geological repository is available for spent nuclear fuel.

The U.S. National Energy Strategy Advocates reviving the growth of nuclear power by standardizing powerplant design, accelerating the introduction of advanced designs, reforming the licensing process, and siting a permanent waste facility.

FUEL CELL TECHNOLOGIES

Additional research and development is needed to produce large fuel cells for bulk electric power systems which ar economically competitive with other generation technologies. Some researchers believe that very high energy efficiencies may be achieved using "molten carbonate" fuel cell technologies. Thus, in future decades, fuel cells may provide a means to use fuels derived from coal in a manner that greatly reduces SOX, NOX, and CO_2 emissions.

RENEWABLE ENERGY RESOURCES

Increased attention is being focused on renewable or "alternative" energy sources as a means of reducing the environmental impacts associated with electricity generation. The technologies receiving the most attention include hydroelectric power, solar power, wind power, geothermal energy, biomass energy, ocean thermal energy conversion (OTEC), and fuelwood energy.

The renewable technologies are being improved to be more economical and practical, and concurrently, increased consideration is being given to including environmental externalities in the analysis of fossil fuel and energy prices. These two factors taken together could potentially hasten the commercialization of renewable energy resources.

HYDRO-ELECTRIC POWER

Hydropower is currently the most extensively used renewable energy resource. Where hydropower resources are being developed, they offer a clean, safe, inexpensive and efficient source of energy.

In the United States, hydroelectric facilities account for approximately 12 percent of operable capacity for electric generation, and provide about 9.9 percent of total net electricity generation. Unfortunately, the development of additional facilities is likely to produce only a relatively small amount of additional capacity. Hydroelectric projects are sometimes opposed for environmental reasons, but it can be shown that many hydroelectric facilities have provided environmental enhancements and recreational opportunities.

GEOTHERMAL ENERGY RESOURCES

After hydropower, geothermal energy makes the second largest contribution as a renewable energy resource in the United States. The Geysers field in northern California supplies 6 percent of California's electric power and accounts for 75 percent of all the installed capacity in the United States.

Efforts are being made by the U.S. Department of Energy to develop techniques for accessing additional geothermal energy. Investigators are searching both for sites like the Geysers (where the earth heats water and turns it into steam) and for other sites where water could be pumped into contact with hot rocks or molten rocks near volcanic areas.

The current outlook for significant energy production from new geothermal sources is unclear. A great deal of heat energy exists near volcanic areas, but the technology to practically access this energy has not yet been demonstrated.

WIND-POWER

In the United States, during the period from 1981 to 1985, the sales of electricity generated by wind-power increased from 21 million dollars to 748 million dollars. This rapid growth was primarily attributable to federal tax incentives for the development of "alternative energy resources," rather than to technological advances. In fact, some of the wind-turbine ventures during this era represented ersatz attempts to adapt helicopter blades, and amounted to very poor applications of state of the art technology with notoriously inferior performance. From 1985 to 1988, given the poor performance of many of the units, and in the wake of federal tax credits having expired, the sales of electricity generated by wind-power dropped from 748 million dollars to 67 million dollars.

Research and development efforts at the Electric Power Research Institute are searching for ways to make wine-power more attractive in terms of both cost and operating performance. One of the technologies being investigated by the electric Power Research Institute involves the use of variable speed turbines. Variable speed turbines offer the advantage of more efficiently converting the mechanical energy resulting from irregular wind speeds and guest into a constant, uniform output in the form of electrical energy.

Even with further technological advances, wind-power is not well suited for areas having relatively low average wind speeds. when wind speed drops to zero or near zero, wind-power is obviously unavailable. Wind-power is generally used as part of a dispatching strategy that recognizes the value of being able to decommit higher cost facilities during periods when wind speeds are sufficiently high to provide "make-up energy." Consistent with this dispatching strategy, wind-power will become more practical if the cost of environmental externalities are increasingly taken into account in calculating the costs of the various energy resources. Wind-power is attractive from the points of view of environmental air quality, water requirements, and minimization of waste products. In some areas, however, environmentalists may oppose wind-power in terms of land use considerations and aesthetics. As in the case of many of the renewable technologies, the economics of using wind-power can be improved if advances can be achieved in energy storage technologies.

Electric utility companies in several states in the U.S. are currently considering adding wind-powered facilities as part of their future generation expansion plans. Electric utility companies in California, Iowa, Minnesota, Vermont, and New York have commitments to install or evaluate wind-powered units.

SOLAR ENERGY

If efforts to reduce capital costs are successful, photovoltaic facilities may provide a very desirable source of peaking capacity and solar thermal facilities may provide a useful source of intermediate capacity. But current solar energy production provides an insignificant contribution to the U.S. electricity grid.

Several solar thermal technologies are being researched including central receiver methods, parabolic solar trough systems, and parabolic dish receivers. Of these only the solar trough systems are commercially available, and in only a few unique situations, are these systems economically competitive.

The current technology to produce photovoltaic cells is rapidly advancing, but currently both concentrator systems and flat plate collector systems are not economically competitive in bulk power markets.

ENERGY STORAGE

Improvements in energy storage technologies can greatly improve the economics of renewable technologies, and can significantly assist in maintain system reliability. The degree to which these benefits will be realized in a given electric system depends on a number of considerations including the composition of the overall system generating mix, the philosophy of dispatching, and the accuracy of load forecasting.

Today pumped storage hydroelectric facilities provide the primary means of storing large amounts of energy for electric generation. Further research and development is needed in the areas of superconducting magnetic energy storage, and compressed air energy storage to reduce the cost and increase the practicality of using these technologies.

T&D EFFICIENCIES

When electric utility companies use higher efficiency transformers or reduce losses in power delivery systems by other means, they reduce both operating costs and the emissions and environmental impacts that attend energy production. Savings in these areas must be weighed against the costs for purchasing and installing new equipment, the waste of discarding existing equipment before the end of its useful life, and the costs for the removal and disposal of existing equipment.

A "lifetime cost analysis" provides a means of making decisions to install new, more efficient, equipment. The useable life span of high efficiency equipment may be somewhat lower than the life span of earlier, lower efficiency, equipment. In addition, assumptions have to be made concerning the possibility that equipment will become less efficient during its life span.

Most U.S. electric utility companies are planning to phase in new high efficiency equipment, such as power transformers, as older equipment is retired and/or when new substations are constructed. This approach is based on the fact that T&D losses using current equipment are fairly low, ranging from 4 to 10 percent. In the developing world, where T&B losses may be as high as 15 percent or higher, the replacement of existing equipment during its useful life may be more easily justified, (34)

Transformers made with amorphous metal cores when used in distribution systems can offer significant energy savings as compared to conventional transformers. Since transformers with amorphous metal cores have lower core losses, they can reduce operating costs, and may have longer lifespans reducing capital costs. (35)

Higher voltage transmission systems can also reduce energy losses. In addition, when a single high voltage line is used to replace multiple lower voltage lines and structures, land use may be reduced.

TECHNOLOGY TRANSFER

Most of the preceding discussion about energy production, delivery. and demand side options has focused on the situation in the United States. Using these options in the developing countries or in those Eastern European countries undergoing industrial modernization could produce even greater energy savings as well as increased environmental benefits.

There is critical need for additional effort in the area of technology transfer. Efforts are needed to share state of the art technical information and efforts are needed to open international markets to companies providing energy efficient technologies. If those from the developed countries establish credibility as having long-term business interest and if financial and institutional barriers can be overcome, it may be possible to avoid "reinventing the wheel" and paying the very significant costs which accompany delayed energy efficiency. The further exploration of these subjects, while important, were considered to be beyond the scope of this paper.

CONCLUSION

Studies performed for the Edison Electric Institute show that modern electrotechnologies, if selectively and intelligently applied, can reduce environmental impacts. Thus, electricity can play an important role in meeting future energy needs.

However, the wide-spread use of energy efficient demand side technologies can be achieved only if it is possible to economically and reliably generate and deliver a sufficient amount of electricity to meet customer needs. Therefore, this paper cautions against focusing only on demand side options. Newly developed technologies and the more effective use of existing technologies can improve the efficiency of electricity production, delivery, and storage.

Finally, it is notable that energy efficiency investments are critically needed in developing countries and in those Eastern European countries undergoing industrial modernization. This paper identifies policy options and advocates technology transfer, but the identification of financial and institutional barriers which may impede these goals were considered to be beyond the scope of this paper.

REFERENCES AND FOOTNOTES

1. DOE/EIA Annual Energy Review, May 1990.

2. Ecowatts: The Clean Switch, Using Electricity to Save Energy and Green House Gases, Science Concepts, April 1991.
3. DOE/EIA Annual Energy Review, May 1990.
4. Impact of Demand Side Management on Future Customer Electricity Demand: An Update, Barakat & Chamberlain, Inc., EPRI CU-6953, September 1990.
5. Saving Energy and Reducing CO_2 with Electricity, Zaininger Engineering Co., Inc., EPRI CU-7440, June 1991.
6. Ibid.
7. Ecowatts.
8. "Energy From Fossil Fuels," Scientific American, September 1990.
9. Handbook of Energy Technology & Economics, 1983, Wiley Interscience, pg. 1036.
10. Energy Efficiency Report, California Energy Commission, October 1990.
11. Ecowatts.
12. Scientific American, August 1987, pg. 10.
13. Ecowatts.
14. American Council for an Energy Efficient Economy letter to U.S. President Bus, December 7, 1990.
15. Saving Energy and Reducing CO_2 with electricity, Zaininger Engineering Co., Inc., EPRI CU-7440, June 1991.
16. Environmental Energy and Economic Effects of Residential Heating and Cooling, Entechnology, Inc., EPRI RP2597-19, Draft Final Report 1991.
17. Handbook of High Efficiency Electric Equipment and Co-Generation Options for Commercial Buildings, EPRI CU-6661, December 1989.
18. Assessment of Gas Industry Competition in the Commercial HVAC Market, EPRI RP2480-2, Draft Final Report, February 1991.
19. Environmental Assessment of the Application of Electric Arc Furnaces to Steel Manufacturing, EPRI Draft Report RP2662-17, March 1991.
20. Environmental Assessment ofthe Application of Induction and Plasma Melting Technologies to the Foundry Industry, EPRI Draft Report RP2662-17, April 1991.
21. Carbon Dioxide Reduction Through electrification of the Industrial and Transportation Sectors, Energy Research Group, Inc., Edison electric Institute Issues and Trends Briefing Paper #54, July 1989.
22. EPRI Technical Brief - Electric Van and Gasoline Van Emissions: A Comparison, October 1989.
23. ERG I.
24. Electric Steelmaking: Recent Trends and Future Constraints, C. C. Burwell, Institute for Energy Analysis, May 1984.
25. Carbon Dioxide Reduction Through Electrification of the Industrial and Transportation Sectors, Energy Research Group Inc., Edison Electric Institute Issues and Trends Briefing Paper #54, July 1989.
26. Ecowatts.
27. Ecowatts.
28. Electricity and Industrial Productivity: A Technical and Economic Perspective, P. Schmidt, Pergamon Press, 1984.
29. Ecowatts.
30. Electricity and Industrial Productivity: A Technical and Economic Perspective, P. Schmidt, Pergamon Press, 1984.
31. Roles of Electricity: Glassmaking, EPRI, September 1986.
32. Ecowatts.
33. Ibid.
34. New Earth 21 Report, MITI Report NR 382, June 1990.

35. High Efficiency Transformers, Correspondence from Allied Signal, September 25, 1991.
36. USCEA - Electricity from Nuclear Energy, September 1991.
37. DOE - Status of the Clean Coal Technology Demonstration Program, September 12, 1991.
38. OTA Report - Energy Technology Choices, August 1, 1991 (Pub. No. 052-003001251-1).
39. EEI Quarterly Statistical Report, September 16, 1991.
40. Global Energy and Electricity Futures, Chauncey Starr, IEEE Power Engineering Review, August 1991.

FOSSIL FUEL AND GREENHOUSE GAS MITIGATION TECHNOLOGIES

Meyer Steinberg

Brookhaven National Laboratory

Within the last two years the world has become well alerted to the Global Greenhouse problem. Certainly the physical science of the potential of the greenhouse effect is quite simple and well understood. When a greenhouse gas such as carbon dioxide, methane, nitrous oxide or chlorofluorocarbons is put into the atmosphere and its concentration increases, the tendency is to trap more of the Earth's surface reflected solar energy in the atmospheric blanket surrounding the earth. This then tends to cause an increase in the temperature of the Earth's surface. However, there are at least three things, that are uncertain about this problem. One is the magnitude and future timing of the temperature increase caused by the greenhouse effect. The second is the effect of this temperature increase on global climate warming, and third is the effect of global climate warming on the terrestrial ecology. In fact there is still some uncertainly as to whether there will be a climate warming or a climate cooling and as a result, the US Department of Energy prefers to refer to the topic as studies in the science of "Global Climate Change," rather to global warming or cooling. Indeed not all climate change is bad. In fact some global warming may be good. Quoting the Russians "global warming will make Siberia that much more livable."

The major greenhouse or radiative gas happens to be CO_2 because its concentration is much higher than the other radiative gases and its concentration has been measured for a longer period of time. Thus the utilization and combustion of fossil fuel natural gas, oil and coal - the major sources of CO_2, are being blamed for the potential greenhouse problem (Table 1). And since, of the three natural fuels, coal emits the greatest amount of CO_2 per

TABLE 1. CRITIQUE OF MITIGATION TECHNOLOGIES FOR REDUCING CO_2 EMISSIONS

Technology	Problem
1. Substitute Natural Gas for Coal and Oil.	1. Limited Resource
2. Substitute Nuclear for Fossil Energy and Solar Energy.	2. Nuclear Fission has Safety and Waste Problems. Solar Energy is Limited Due to Storage and Low Intensity.
3. Remove, Recover, and Dispose CO_2. In Ocean. In Depleted Wells.	3. Doubles the Cost of Power. Unknown Ocean Ecology. Depleted Wells have Limited Capacity.
4. Convert CO_2 to Consumer Products.	4. Limited Markets for CO_2, Capacity Mismatch.
5. Energy Technologies for Improved Efficiency and Conservation.	5. Reduces CO_2 Emissions and Yields Return on Investment Due to Fuel Savings.

Global Enery Strategies: Living with Restricted Greenhouse Gas Emissions, Edited by J.C. White, Plenum Press, New York, 1994

TABLE 2. CLEAN COAL TECHNOLOGY PRODUCTION OF A CLEAN CARBON FUEL AND CO-PRODUCTS

HYDROCARB PROCESS

All Coal Ranks ←------------- Recycle ------------- Carbon Black Main Product

$C\ (Impure) \rightarrow H_2 \dashrightarrow CH_4 \dashrightarrow H_2 + C\ (Pure)$

EXOTHERMIC   ENDOTHERMIC

Optional Additions	Co-Products	
CO_3 - Limestone	----→ H_2	- Hydrogen
H_2O - Water	----→ CH_4	- Methane, SNG
CO_2 - Carbon Dioxide	----→ CH_3OH	- Methanol
Alt. Feedstocks	**Waste Streams**	
Peat	---→ H_2O	- Water
Wood	---→ CO_2	- Carbon Dioxide
Rubber	---→ $CASO_4$ or S	- Sulfur
Paper	---→ N_2	- Nitrogen
MSW	---→ Ash	

Three Methods of Heat Transfer - Gas, Solid, or Steam

Three Reactor Types - Fluidize Bed, Moving Bed, Entrained Bed

unit of thermal energy and since the resource of coal is at least an order of magnitude larger than oil or gas, coal is being singled out as the major culprit causing the future greenhouse global warming problem. So everyone is "beating up on coal."

However, most of the money being spent by government agencies on the greenhouse goes toward understanding the science of global climate change and very little funding goes into directly studying what can be done about the greenhouse. One of the main themes at a recent meeting on geoengineering for mitigating the greenhouse problem was to reduce the utilization of coal. The nuclear people were saying "stop burning coal." They say they can supply energy without CO_2 generation. They see the global greenhouse as a God-send for rejuvenating the use of Nuclear Energy. On the other hand, the Solar people are saying, "FIE on both your houses." They say, "we don't need coal and we don't need nuclear. We can generate the needed energy only from the sun." The fact remains that we will need to use some or all of these energy sources depending on the local conditions. However, one thing is certain, the world cannot afford not to use its main energy source and that is coal.

Besides adapting to climate change the coal industry needs to structure itself so as to conceive of and adopt technologies that will respond to the challenge of utilizing coal, in spite of the greenhouse problem. This boils down to devising technologies that will utilize coal with reduced CO_2 emissions. We have been studying this problem for over 10 years and have made reviews and assessments of CO_2 reducing technologies.[1] These include, improved energy utilizations efficiency by conservation, post combustion, removal and recovery of CO_2 from power plant stacks, and disposal of CO_2 in the earth and the ocean. Of course, there is one axiom concerning the factoring in of all external costs for protecting the environment. That is, "no matter what process is being considered or product being produced, there will always be an increase in economic cost." The problem is to minimize this cost while achieving environmental stability. Here, however, in this paper, I would like to review another option for the coal industry. This has to do with precombustion treatment of coal by a process called HYDROCARB. There have been several stages in the development of this process, which I would like to now review (Table 2).

The process was conceived at Brookhaven National Laboratory for the purpose of producing a clean fuel from coal to overcome the disadvantage of coal in competing with oil and gas in the heat engine market.[2] The main disadvantage of coal, in this respect, is the ash and sulfur content which eliminates coal from being utilized in conventional turbine and diesel engines. We recognized that it is economically impossible to simulate a hydrocarbon such as natural gas and gasoline produced from hydrogen deficient coal, as long as low cost crude oil is available from the ground at costs of less than $15 to $25 per barrel. This has become evident after the demise of the Synthetic Fuel Corporation when the price of oil dropped from above $40 per barrel during the oil embargo days to less than $20 per barrel in the last several years. We thus devised the HYDROCARB process which goes in the opposite direction. It produces pure carbon as the main product and hydrogen-rich coproducts consisting of either the gases hydrogen or methane, or liquids either methanol or gasoline.[3.9] In other words, the coal, which also contains hydrogen and oxygen in addition to carbon is essentially cracked to its elements, mainly carbon and hydrogen and to compounds which contain oxygen such as water and methanol (Table 3).

In the preferred process mode, (Figure 1) there are three reaction steps: (1) hydrogenation of coal to methane and CO, (2) conversion of CO with hydrogen from coal to methanol and (3) the thermal decomposition of the methane to pure carbon (carbon black).[5] The key to the economics of the process is the recycling of the hydrogen. In contrast to the conventional process of producing a hydrogen-rich synthetic or substitute natural gas or gasoline fuel, which requires an outside source of hydrogen, the HYDROCARB process produces a carbon-rich fuel and a hydrogen-rich coproduct, gaseous or liquid fuel, relying only on the composition of the fuel feedstock itself. Furthermore, the process efficiency is much higher than conventional synthetic fuel processes. We believe this process can compete with the current price of fuel in the range of $15 to $25 per barrel of fuel oil equivalent (FOE) depending on the cost and type of coal feedstock. We are recommending building large mine mouth coal refineries to produce clean carbon and methanol and, depending on the acceptance by the automotive industry, the methanol can be further converted to gasoline.[7]

What does this have to do with the greenhouse? First, in converting coal to a synthetic fuel the HYDROCARB process does not generate any CO_2, unlike conventional synthetic fuel

TABLE 3. GENERATING ENERGY FROM FOSSIL FUEL WITHOUT CO_2 EMISSIONS

Principle:
- Extract Hydrogen from Fossil Fuels.
- Use Hydrogen only as Fuel.
- Return Clean Carbon to Ground for Possible Future use as CO_2 Environment Permits

Fossil Fuel	Ht. of Combustion HHV Kcal/Mol	Cracking Process	Atomic H Fraction in Fossil Fuel	Heat of Cracking Kcal/Mol-C	Net Energy in Hydrogen % of Fossil Fuel
Nat. Gas	-212	$CH_4 \longrightarrow C + 2H_2$	80	+18	64 - 5 = 56%
Petroleum					
Alkanes	-165	$CH_2 \longrightarrow C + H_2$	67	+6	41 - 4 = 37%
Aromatics	-142	$CH \longrightarrow C + 1/2H_2$	50	-3	24 + 1 = 25%
Coal	-116	$CH_{0.8}O_{0.08} \longrightarrow C + 0.08\ H_2O + 0.32\ H_2$	43	0	19 ± 0 = 19%

NOTE: Ht. of Comb. - of C = -94 Kcal/Mol.
HHV of Comb. - of H_2 = -68 Kcal/Mol.

processes. Secondly, it is more efficient than conventional synthetic fuel processes, so that, overall, it generates less CO_2 per million BTU of energy generated. Third, we are dividing coal into a carbon-rich fraction and a hydrogen-rich fraction. If we are concerned about the greenhouse, store the carbon and utilize only the hydrogen-rich fraction. Because of its physical properties, it is much easier and less risky to store solid non-volatile carbon black at the mine than it is to dispose of CO_2 gas after combustion in an uncertain way, either in the ocean or in depleted gas wells or in natural void spaces in the earth. Of course there is a penalty to pay, because only up to 24% of the energy in coal can be utilized as hydrogen without any CO_2 emission. Burning coproduct methanol while sequestering the carbon could increase the coal energy utilization to 40% while reducing CO_2 emissions by 45% compared to directly burning the coal. Furthermore, should the greenhouse effect not become severe in the future, the carbon can be taken out of storage from the storage mines and used as fuel, and this time the carbon would be clean, since the ash and sulfur would have been previously removed.

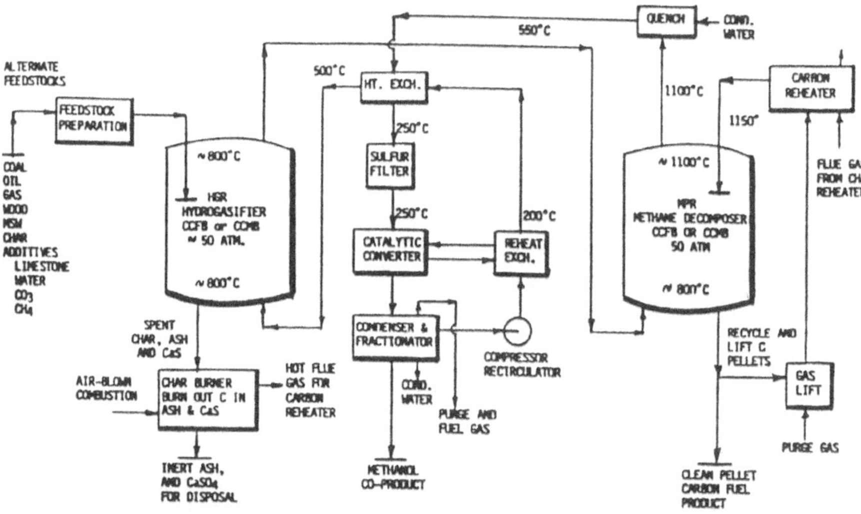

FIGURE 1. CARBEX VERSION OF THE HYDROCARB PROCESS CLEAN CARBON AND CO-PRODUCTION FUELS FROM CARBONACEOUS FEEDSTOCKS

By coprocessing coal with biomass by the HYDROCARB process (Figure 2 & 3) and utilizing the hydrogen-rich fraction while storing the carbon, the CO_2 emissions would be reduced to negligible values while increasing the energy utilizing of coal beyond that of HYDROCARB processing coal alone.[4] In this case, we are utilizing free solar energy to remove CO_2 from the atmosphere by growing trees, thus taking advantage of the natural photosynthesis process and combining this with coal to reduce CO_2 while producing a synthetic clean methanol or gasoline fuel.[3,9] In fact, coprocessing natural gas or oil with biomass can actually yield a negative emission of CO_2 which means a reduction of atmospheric CO_2 when storing the carbon and utilizing only the methanol or gasoline product fuel (Table 5).

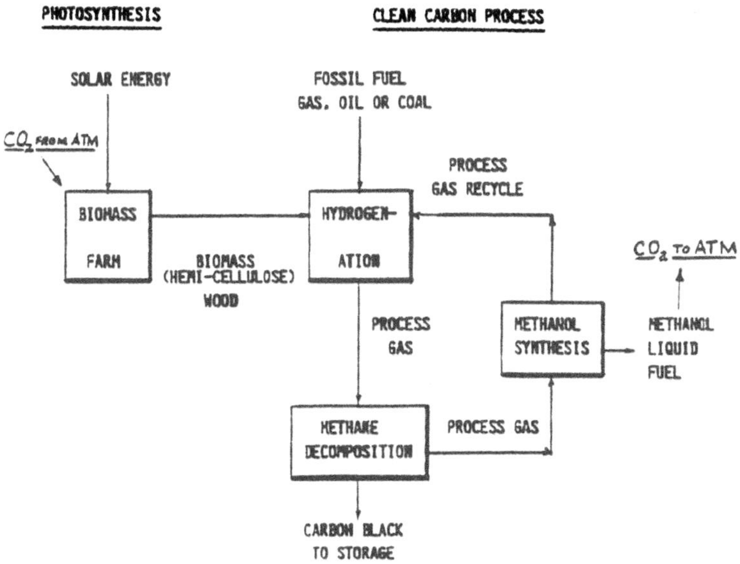

FIGURE 2. CO-PROCESSING BIOMASS WITH FOSSIL FUEL VIA THE CLEAN CARBON PROCESS FOR REMOVAL OF ATMOSPHERIC CO_2 AND PRODUCTION OF METHANOL

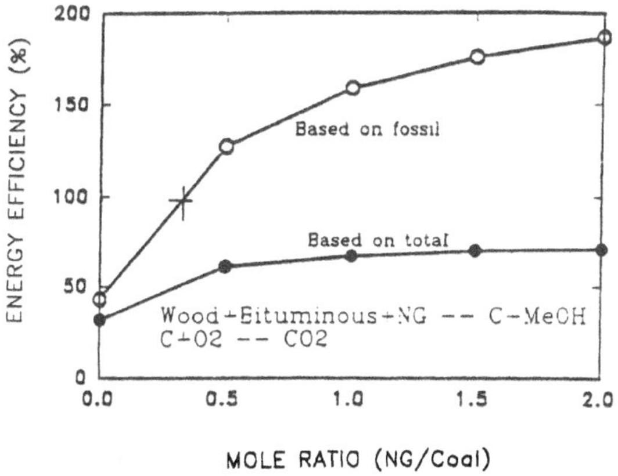

WOOD COAL

A - NG) $0.32\ CH_{1.44}0.66 + CH_{0.8}O_{0.1} = C + 0.32\ CH_3OH$

NG

B - C) $1.0\ CH_4O_{0.66} + 0.3\ CH_4 = 0.64\ C + 0.66\ CH_3OH$

C Burn C $C + O_2 = CO_2$

For 100% Fossil Fuel Efficiency
CH_4/Coal Ratio = 0.3

Approximately Equal to Recoverable Reserves

FIGURE 3. COPROCESSING COAL AND NATURAL GAS WITH BIOMASS FOR ZERO CO_2 EMISSIONS

By further coprocessing coal with oil or gas and biomass an additional increase in utilization of coal with reduction in CO_2 emission can be realized. In this case, what we are recommending is that before we deplete all of our golden fuels, oil and gas, we use these high hydrogen content resources to extend our vastly greater resource of coal to produce hydrogen-rich coal derived fuels, while storing less coal-derived carbon to reduce CO_2. In this way, all of our fossil fuel, reserves as well as biomass resources will be most efficiently utilized. The key to these procombustion cleaning systems is the HYDROCARB process which produces a clean carbon fuel and methanol or gasoline fuel from all carbonaceous raw materials.

A final dimension for this technology. Whereas most post combustion, removal, recovery and CO_2 disposal schemes are only applicable to large central power stations, the HYDROCARB process supplying coal derived methanol and reconstituted gasoline product fuel can be used in all sectors of the energy economy, the utility, industry and the transportation markets, in an environmentally acceptable manner. The process is thus worthy of further development.

It appears that the State of Alaska is interested in undertaking the construction of an integrated HYDROCARB Process Demonstration Unit looking towards full scale commercialization of the process for developing the vast coal resources in the State of Alaska as well as other worldwide coal resources. The Environmental Protection Agency together with the South Coast Air Quality Management District is undertaking to build a pilot plant disposing of biomass and sewage sludge by coprocessing with natural gas, employing the HYDROCARB process.

REFERENCES

(1)M. Steinberg, J. Lee and S. Morris, "An Assessment of CO_2 Greenhouse Gas Mitigation Technologies," BNL 46045, Bookhaven National Laboratory, Upton, NY (March 1991).

(2)E.W. Grohse and M. Steinberg, "Economical Clean Carbon and Gaseous Fuels from Coal and Other Carbonaceous Materials," BNL 40485, Brookhaven National Laboratory, Upton, NY (November 1987).

(3)M. Steinberg, "Coal to Methanal to Gasoline by the HYDROCARB Process," BNL 43555, Brookhaven National Laboratory, Upton, NY (August 1989).

(4)M. Steinberg, "Biomass and HYDROCARB Technology for Removal of Atmospheric CO_2," BNL 44410, Brookhaven National Laboratory, Upton, NY (February 1991).

(5)M. Steinberg and E.W. Grohse, "Production of A Clean Carbon Fuel and Reproduct Gasous and Liquid Fuels from Coal by the HYDROCARB Process System," BNL 46490, Brookhaven National Laboratory, Upton, NY (November 1990).

(6)R.H. Borgevardt, M. Steinberg, E.W. Grohse and Y. Tung, "Biomass and Fossil Fuel to Methanol and Carbon Via the HYDROCARB Process," EPA Paper In Press, (March 1991).

(7)M. Steinberg and E.W. Grohse, "Economical Clean Carbon Fuel and Co-product Gaseous and Liquid Fuel from Coal," BNL 42489, Brookhaven National Laboratory, Upton, NY (September 1989).

(8)M. Steinberg, E.W. Grohse and Y. Tung, "A Feasibility Study for the Co-processing of Fossil Fuel with Biomass by the HYDROCARB Process," BNL 46058, Brookhaven National Laboratory, Upton, NY (April 1991).

(9)M. Steinberg and E.W. Grohse, "Hydrocarb-M^{SM} Process for Conversion of Coals to a Carbon-Methanol Liquid Fuel (Carboline-M^{TM})," BNL 43569, Brookhaven National Laboratory, Upton, NY (January 1989).

THE U.S. ENERGY STRATEGY

Edward R. Williams

U.S. Department of Energy

In this paper I discuss the national energy strategy in relation to forecasting the changing path of greenhouse gas emissions.

We have begun the first meetings of the inter-governmental negotiating committee, preparing for the time when the environment convention will go into place on global climate change. If you recall, in June of 1991, President Bush made a challenge to the developed countries to provide the first copies of their national action plans early in the process, and promised that the United States would have theirs together before the end of the year. It was distributed in Geneva in December, 1992.

A major part of that plan is the national energy strategy as it's being carried out with the Environmental Policy Act of 1992, the act that was signed by President Bush on October 25, 1992.

There are many other things in the action plan as well, but I will speak specifically to those areas that deal with the national action plan, not the voluntary programs such as EPA has in the green area and that type of thing, nor those areas that deal with agriculture and the emissions associated with that. I shall deal with the original national energy strategy as well as how that has changed as the energy policy act went into place. There has not been a major change, at least in the gross figures.

The national energy strategy relies primarily on advanced technologies and improved energy management practices to maintain an adequate supply of energy while maintaining the environment. In the national energy strategy report, with many of the pollutants, there are improvements just because of the national energy strategy itself. The same holds true with the greenhouse gases. It achieves these reductions because of reduced energy consumption. It also improves energy and use efficiency and includes fuel switching, especially in the early years.

But a major part of the strategy is what will happen in the long term, what will happen as new energy technologies can be developed and put into place. The national energy strategy as it's reflected in the national action plan on global climate change is a case of the U.S. taking prudent strategies to reduce greenhouse gases. As we said as early as 1989, these are strategies that are justified on grounds other than global climate change, but they do improve the emissions profiles and forecast for the future.

In the national action plan, you will find a litany of small actions that are associated with the national energy strategy. Connected with transportation, we have a number of things that deal with alternate fuel vehicles, programs that were started under the Clean Air Act and will be intensified under the energy policy act. Also the act talks about fuel switching, going

Global Enery Strategies: Living with Restricted Greenhouse Gas Emissions, Edited by J.C. White, Plenum Press, New York, 1994

from gasoline to biomass, and some uses of natural gas; allowing the fuel switching from some of the fossil fuels which have higher intensity of carbon release. Nuclear also plays a major role in the actions that are included in the strategy and now have been ratified by statutory actions within the policy act itself.

In the forecast, a major element is, what model was used and what assumptions were used. The model used is one that's been at the Department of Energy since its inception. It's now called Fossil Two; next year it will have a new name, called Ideas. Essentially, we have demand and supply sectors, linked by the energy price feedback.

We represent the energy technologies in quite a bit of detail. About three years ago, the model was extended from looking at the time period of about 1980 to 2000, to look beyond that to the year 2030. This is important when we talk about energy, because there are very few energy R&D strategies that will deliver in the 2000 year time frame. We're talking about longterm introduction of new energy supply technologies. Many of those will not show up until well into the 21st century.

The crux of the matter is usually the assumptions that are used in the model. We show where we're going, and must give a status on where we are. We call it the current policy case. It was the current policy in November of 1990, which at that time talked about improving economic efficiency by use of the markets. It had economic growth rates which I suggest are quite optimistic, but they were the economic growth rates of the Administration, reflected in the national energy strategy and in the reports that supported it. World oil prices certainly had a major impact.

The current policy base case assumptions are:

1. Free market - market choices promote economic efficiency

2. Economic growth rates 2.9%/yr. (1990-2010; 1.8%/yr. (2010-2030)

3. World oil prices 4.0%/yr. (1990-2010); 1.0%/yr. (2010-2030)

4. Technology assumptions "cautiously optimistic"

5. Energy efficiency improvements driven by increasing prices: overall economy becomes 20% more efficient by 2010

6. Oil and gas reserves and resources:
   oil 80 billion barrels
   gas 700 trillion cubic feet
   no access to restricted areas (e.g., ANWR)

7. Alternative transportation fuels: no significant penetration without policy changes

8. Nuclear power: no new nuclear plant orders; no life extension

9. Clean coal technologies: penetration of demonstrated technologies due to favorable costs

10. Renewable energy: steady improvements in renewable electric technologies

The technology assumptions were "Cautiously optimistic." What does that mean? We considered the resources, and the conditions without the new national energy strategy. For transportation we stayed with the use of gasoline, with little penetration of alternate fuels. We anticipated that nuclear power would remain essentially level with where it is today to

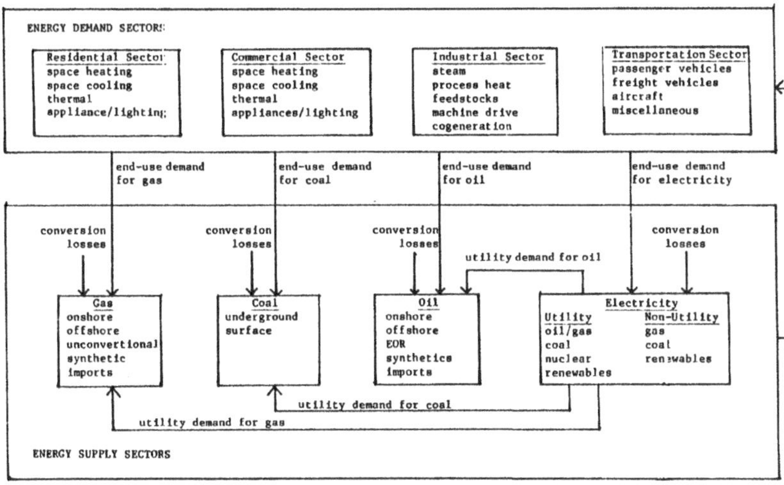

FIGURE 1. THE FOSSIL 2 U.S. ENERGY MODEL

2010, and then start phasing out. There would be no replacement. Clean coal technologies would penetrate over a period of time, but don't look for any of these until well after the year 2000. Renewable energy has steady improvements, but steady improvements from a very small base. We're talking about perhaps four quads today. In the year 2030 it might grow about nine quads, not a big increase, but it would be coming on.

Let's consider the economic assumptions in the projections for the national energy strategy, when we varied how fast GNP will grow and the world oil price. It makes quite a difference. We're only talking about 20 years, but if you have a high GNP growth emissions would grow rapidly. If you assume a lower growth rate, emissions are much less. The price of oil impacts all productions for future emissions because they influence GNP.

With the new energy legislation, we expect demand will decrease from the current policy case, because we include a number of programs which will reduce the demand. In particular, the alternate fuels and transportation research and development will cut down on the amount of oil consumption. We will have a decrease in natural gas, even though we'll have some fuel switching from coal to natural gas. The reason for this is that renewable energy will replace natural gas faster than natural gas will replace coal. Renewable energy will be up.

On primary production, there will be some increase over the 20 years. It comes about because of R&D in renewables, advanced oil production, nuclear beginning to deliver and therefore turning us from a situation where nuclear would be phased out to a situation that by the year 2030 we would have 50 percent more than we have now. Not a very large growth, but definitely growth, and reversal of phasing out. The need for electricity will decrease because of the IRP programs and demand side management.

What are the impacts on the global climate? Since we have the Montreal protocol, with reduced CFC's and less other greenhouse gases, the national energy strategy would mean that we would never have the same impact on global warming from our emissions in the future that we had in 1990.

These predictions are under the old model. A year later, everyone said, well, this is very nice, but the GMPs for the Montreal protocol are probably wrong, and after a small decrease due to the Montreal protocol and due to efficiency, we would have steady growth in greenhouse gases.

In the path that we were discussing we included national energy strategy and the other activities. There will be steady growth in CO_2 after some fuel switching and some efficiency improvement in the short term. The national energy strategy will moderate all the way up to about 2015. This is because of fuel switching, and because of improvements in efficiency. It doesn't have much to do with the introduction of new energy technologies. But those new energy technologies will show up later, when CO_2 essentially is stabilized.

What caused it, when we did the strategy? A number of NES actions were considered; clean coal technologies, energy efficiency standards for federal building, and transportation energy efficiency R&D.

Which ones had the major impact? In the year 2000, the integrated resource planning or the demand side management in electricity played a major role, a 25 percent reduction. Alternate fuels, much the same. Natural gas reform was next, municipal solid waste, next.

What was it worth? It's worth about 200 million tons of CO_2 per year, a pretty small number,as compared to the other values. In the year 2030, it will grow to be about ten times that. It has stabilized at a higher level. What shows up? The research work that goes on in transportation continues to play a major role, but the biggest role is played by nuclear reform. It could be nuclear reform or it could be renewables, but the issue is, how do you switch to no emissions technologies within the market economy? How will that be stimulated if you do want to decrease the gases as quickly as possible? Efficiency still plays a major role. In fact, industrial efficiency, plays the larger role and you have a number of smaller actions.

As we go into the Clean Air Act and the National Energy Strategy, we see little or no change in carbon emissions up to 2010, compared to the present course of events. This is because the R&D has not yet become effective - it will show greater effects later. The national energy strategy contributes to the overall national action plan, but their are many other influences as well, many of them voluntary.

You will see qualitative statements that deal with what state and local governments are doing. They too are improving. In the national action plan, however, is the first assessment of what we believe would be done under the so-called no-regrets policy, an indication of qualitative areas and where that may improve as we do further assessment. This would represent the NES and its impact on emissions, and from that the translation to the impact on global climate change.

INTEGRATING ENERGY AND THE ENVIRONMENT

Alex Cristofaro

United States Environmental Protection Agency

I would like to discuss EPA's agenda at this point in the climate change area. It's not an agenda that is set in concrete.

We have an agenda that is both domestic and international. As you probably know, the U.S. accounts for about 20 percent of carbon dioxide emissions globally, and we are the largest source of carbon dioxide in the world. But while our share is large now, most of the growth in greenhouse gas emissions is projected to occur in developing countries. So our agenda at EPA is to try to address both, an international focus as well as a domestic focus.

First, let me tell you a little bit about our international strategy. What we would like to do is to try to make the climate convention work, and specifically make the national planning process within the climate convention work. The plans requires that countries try to develop national plans with an aim to returning to 1990 levels. In this planning process, they are able to take into account their economic conditions, their starting points and so on, but they do have to develop plans that aim to return to 1990 levels.

We're hoping to promote environmentally friendly passive development in developing countries, where most of the growth in climate change emissions or greenhouse gas emissions is projected to occur. We expect that in the next ten years, developing countries could account for as much as half of greenhouse gases in the world.

We have been very active in supporting forest initiatives at the EPA, which is a very new area for us. President Bush at Rio announced a $150 million Forests for the Future initiative, and EPA has been co-leading this effort. We're hoping to use this effort in part as a joint implementation of the climate change convention. We're also hoping to keep the international process, in particular the IPPC, focused on the right things so that we can move forward in the area of protocols.

What can we do to make the convention work? First, we can assert U.S. leadership and demonstrate that we take the convention seriously. Second, we are engaged in a country studies program in which we are providing $25 million to developing countries to assist them in their national planning process. Third, we're hoping to have what's called joint implementation, where countries cooperate to meet the goals of the framework convention. Fourth, we're hoping to develop the basic technical infrastructure that is necessary to implement any kind of climate change agreement. It is necessary to start at ground level in many countries and develop things like emission inventories to calculate what the emissions are. In many countries there is very little information, and we're working through the OECD to help improve the data and the basic technical infrastructure that is necessary to develop national plans in developing countries.

In the area of technical assistance to developing countries, as I mentioned before, we have a number of country studies. We try to affect the policies of multilateral development banks like the World Bank where we've had a few successes in getting them to support training programs for demand side management in developing countries. We would like to have more environmentally friendly structuring of the global environmental facility, to provide technical assistance to developing countries. We're working through bilateral agreements with a number of countries, in the area of climate change activities, in pilot projects for different kinds of technologies, and in forestry projects. We provide direct technical assistance and we're hoping through our efforts to stimulate cleaner growth through policy changes at the macro level.

Energy subsidies are rather large in the globe right now. The World Bank has calculated, for example, that there are $240 billion energy subsidies in the world, and were it not for these subsidies, global carbon emissions would be ten percent lower. Thus there's a whole host of policies, both macro policies as well as technology demonstration projects that we think make sense. We also have been working on the forestry initiative that the Bush Administration has started, our administrator is co-chairman of an inter-agency task force to implement this $150 million program. As with all programs and all causes, the longest journey begins with a single step. The single step in our 1993 budget is a five million dollar program within EPA that we're hoping will grow in the next appropriation. We're working specifically with the countries of Russia and Mexico, trying to design forestry projects that can both serve the goals of protecting forests and provide sources of emission offsets for companies in the U.S. that have policies that require them to seek emission offsets when they want to build new electric capacity.

With respect to the domestic agenda, there are a number of options that could be pursued. In EPA, we believe that a balanced portfolio of approaches should be under consideration. We don't think that there's any one alternative that should be pursued uniquely. The approaches are volunteerism and market creation, which are means for creating marketable permit systems for carbon dioxide and tax and subsidy programs. We also can consider state programs and state electric utility regulation, which can have a very large effect on greenhouse gas emissions. There are also the federal government programs in the agriculture area, in the transportation area and the natural resource area that will have a very large effect on emissions. The old standby is actually regulating, an option that could be pursued with all the others.

We've all seen in the action agenda that we have at EPA a number of voluntary programs, but our program right now is focused on technologies. We go out to companies and we tell them that it would be in their interest to sign up to a program to evaluate their lighting systems or their motors, and to change them where they can be profitable. We've had a tremendous response to this Green Lights program. We've had 600 participants.

In addition, we have another program, called the Energy Star program, which has been successful in the area of computers. EPA people went to computer companies and said, did you know that computers run a very large number of hours during the day, even when people aren't using them? If you could have a way of making the computer sleep while people haven't touched the keys, that you could save a lot of energy. The response has been very positive, and a number of companies that produce computer chips have agreed to change computer systems so that they will sleep when they're not in use.

Through our voluntary programs in the action agenda, we had projected that we could save 45 to 75 million tons of carbon or carbon equivalents. In 1991, we've had a total of $14 million in expenditures due to our Green Lights program, seven million of which was due to electric utility re-rate programs. Relative to the total amount of national lighting expenditures, it's not a whole lot. It's less than half of a percent, but the program is expected to grow over time.

With regards to programs such as Green Lights, Green Motors, Green Computers, there's no way that you can capture the full breadth of energy efficiency opportunities in the economy by focusing on technology by technology. Many of the technologies covered by the voluntary programs we stress will have energy efficiency standards at some point in the near future.

There's a great deal of energy use that is not covered by our voluntary programs or by energy efficiency standards. That is a large source of untapped potential for volunteerism and we're trying to design additional voluntary programs to cover these areas.

The amount of carbon that it takes to produce a dollar's worth of GNP, is 40 percent less now compared to 1950. Through time in any economy, there is technological change that produces efficiency improvements. The question with volunteerism in my mind is whether or not you're just seeing people taking credit for a trend that's occurring anyway, or whether there's real effort underway.

But clearly, one way to accelerate these trends is by providing market incentives, such as taxes to steer people towards energy efficient activities. Taxes are definitely a real alternative. There are a number of different forms they can take. You could tax carbon, you could have an excise tax on energy or on fossil fuels. You could tax on the basis of their BTU content. You could try to tax things like gasoline or you could do any kind of combination. If you choose a tax strategy to address climate change or to address greenhouse gas mitigation it's strongly suggested that the revenues be recycled, because this will not only reduce emissions, but it can stimulate economic growth.

Death and taxes may be inevitable, but if you're going to have to have taxes, it may be better to tax bad things rather than good things. If you don't tax energy, then you'll be taxing capital or labor. So either you provide an incentive to use less energy, or you're providing an incentive to use less capital or to work less. Those latter two options have negative consequences for the economy.

Our studies at EPA basically involve looking at carbon taxes. There's a second problem with carbon taxes, in that the revenues that you raise with carbon taxes are so large or can be so large that you have to do something with them. You have to either retire the deficit or change your fiscal policy. Then you start thinking about how you would change the rest of the tax structure.

If you use the revenues to provide an investment tax credit, then carbon taxes could actually stimulate economic growth. If you use them to reduce business taxes which are taxes on capital, then the economy also grows. Alternatively, if you use them to reduce personal income taxes, then there's the negative consequences for the economy. So the effect of carbon taxes on economic growth depends largely on what other fiscal changes you make in conjunction with carbon taxes.

If you want to raise revenue through environmental taxes, what kind of taxes should be on the table, and what are the effects? If we have a BTU tax that is designed to raise $20 billion, by the year 2000, it would yield about a 18 million ton reduction, and in the year 2010, that would grow to about 27 million tons.

There are several subsidies that affect greenhouse gas emissions in the U.S., and we have been trying to take a look at some of these. Changing those could also reduce greenhouse gas emissions.

Untaxed perks are such a subsidy. Many people receive parking spaces from their employers which is a benefit that is not taxed. You could conceivably change the tax treatment of employer provided parking. As I recall, the World Resource Institute calculated that that was an $80 billion tax expenditure.

Another thing that you could do is to try to impose more user fees. For example, state maintenance of highways is often funded out of general revenues, and gasoline taxes are insufficient to maintain state roads. There is an implicit subsidy that's provided there. There's the old standbys. You could reduce below-cost timber sales. You could reduce agricultural subsidies. Those would both reduce greenhouse gases. You could reform federal power marketing pricing policy, where hydroelectric power from federal dams is provided and is averaged in with higher cost forms of generation, a market distortion caused by the federal government. There are certain areas in fossil fuel production, tax exemptions and so on, that could be examined.

As to other forms of programs, there's offset and training programs, the Cooper-Sinar approach. Some of the characteristics of these programs are that they tend to focus only on electric utilities or only on large sources, so they tend to be piecemeal solutions to this problem. The second thing about them is that if you focus exclusively on new sources, saying that any new generation has got to obtain the mission offsets, old sources stay old. People are less likely to purchase new sources and are more likely to keep their old sources going for a longer period of time. When we looked at these types of programs, we found that they don't produce any kind of emission reductions for a good ten years or so, because people are able to avoid the programs or to avoid the incentives that are envisioned in the programs by operating their existing capacity for greater periods of time.

There are some electric utilities -- three or four of them, that are voluntarily signing up to offset any new emissions that they generate. Some are New England Electric, PG&E, and SoCal Ed.

We also have a focus in EPA on automobiles. They are a source of greenhouse gas emissions, and the options to address autos are either CAFE tightening or some other kind of programs such as super-guzzler rebate programs. We have a preference for super-guzzler programs, because they're more likely in our view to result in peoples' selection of the size of automobiles. The National Academy of Sciences says that 32 miles to the gallon is technologically feasible as an option. We've done some studies that indicate that you can maybe go somewhat higher than that. However, a super-guzzler rebate program, is a revenue neutral kind of program where for every car that gets more than 27 miles per gallon receives a subsidy for every mile per gallon. Cars below 27 miles per gallon are taxed a certain amount. People then, because they see a market signal, are more likely to pick cars that are smaller and more fuel efficient, which is not something that you necessarily get with a pure CAFE program.

In any kind of CAFE or automobile targeted program, there is also the ability to try to lessen the impacts on the companies by providing investment tax credits and to lessen the impact on safety by increasing safety standards.

In terms of regulatory options, there are many things that people don't usually think of. For example, landfills emit large quantities of methane. The EPA right now is in the process of issuing a regulation to control volatile organic compound emissions from landfills, but there will be an ancillary benefit of methane control as we go through this. We proposed this regulation about a year and a half ago, and we're at the point now where we're ready to finalize it. We could get greater emission reductions from this regulation if we target smaller and smaller landfills. But the problem there is that the number of regulated sources goes up and the number of small communities that would be impacted by our regulation also goes up. Small communities have been hit with many EPA regulations of late, for drinking water, landfill and solid waste disposal, so there's a question here of whether or not the tradeoff is worthwhile, the emissions capture versus the hardship that may be caused by the program.

Another area of attention within the EPA is the area of forest. To put forest in perspective, I think it's an area that is often overlooked in discussions about greenhouse gases. There tends to be a focus only on energy. But our total emissions in the United States of carbon dioxide are about 1.3 billion tons, and our national forests in the United States currently sequester 34 billion tons of carbon. The inflows and outflows of carbon in our forests are rather large. In 1990, 284 million tons of carbon flowed into the forests, 167 million tons flowed out, resulting in a net storage of 117 million tons. The scale is such that forest and forest options should not be overlooked.

We have been looking at a number of alternatives in the EPA. Just to give you a feel for what's on the menu, you can reduce the amount of timber that's harvested from national forests. You can engage in tree planting programs. You can encourage paper recycling. There was actually a recent interesting proposal in the Progressive Policy Institute with respect to paper recycling, in which we would assign recycling targets or recyclable content targets for newsprint, and then give users the opportunity to either meet those targets or to buy credits from other people who can recycle paper. That's an interesting option that I think we'll probably be analyzing in the future.

You can plant on Department of Defense bases, and you can change the way farmers manage soil. Finally, in the area of transportation, we're working with the Department of Transportation and with state and local agencies on the implementation of the highway bill passed last year, which provides $140 billion, six billion of which is supposed to be spent on air quality management and congestion programs. We've been working with the Department of Transportation, which is required by law to consult with us on the implementation of this legislation. We are attempting to figure out what is the best way to spend this money to improve air quality and to reduce greenhouse gas emissions. If you spend highway money to increase the number of highways, that obviously is going to have a negative consequence for greenhouse gases. But if you can design some programs to either purchase alternative fueled vehicles, which is something that the Denver municipality has recently contacted us about, or to design innovative systems for highway pricing, then you can reduce emissions. It's a big area.

I am pleased to have been able to lay out what's on our agenda at this point in time.

REDUCING U.S. CO_2 EMISSIONS - THE VALUE OF FLEXIBILITY IN TIMING

Alan S. Manne

Stanford University

and

Richard G. Richels

Electric Power Research Institute

INTRODUCTION

Within the international community there has developed a growing consensus that greenhouse gas emissions should be limited, and the debate has shifted to the issue of targets and timetables. Who should reduce emissions? By how much? And how quickly? The European Community (E.C.) is proposing that industrialized nations agree to stabilize CO_2 emissions at 1990 levels by 2000. This is seen as a first step in a longer term strategy towards more substantial emission reductions. Alternatively, the U.S. is advocating a more flexible approach, one that would provide more freedom in managing the transition to a less carbon intensive economy.

The choice of a strategy depends, at least in part, on one's estimate of the economy-wide impacts. Proponents of the E.C. proposal suggest that the targets and timetables could be achieved at little or no economic loss. The U.S., on the other hand, is considerably less sanguine. For example, The U.S. Department of Energy suggests that a tax of the order of \$100 per ton of carbon would be required in 2000 in order to stabilize CO_2 emissions at 1990 levels (Bradley et al., 1991).

The costs of a carbon limit are sensitive to a number of factors. In a previous study, we examined how the costs depend on the supply technologies and resources as well as on the demands themselves (Manne and Richels, 1992). Plausible assumptions were found to support a wide range of cost estimates. Nevertheless, the analysis suggested that stabilizing emissions by the turn of the century will entail sizable losses for the U.S. economy.

In this paper, we extend our earlier work to the issue of timing. One reason why compliance costs tend to be substantial is the shortage of low-cost substitutes during the early decades of the 21st century. There are constraints on the rate at which new supply and end-use technologies can enter the marketplace. Suppose that the U.S. has more time to manage the transition away from fossil fuels. What would be the implications for costs? What would be the implications for cumulative emissions?

As a starting point, we examine the E.C. proposal and its potential impact on U.S. GDP (gross domestic product). We then explore the implications of slipping the proposed timetables by ten and twenty years. The analysis has required us to make assumptions about post-2000 emission limits - a subject that has received remarkably little attention in the recent debates. Compliance costs depend not only on the year 2000, but also on what is expected thereafter. We examine two scenarios: one in which emissions are ultimately reduced by 20% below 1990 levels and one in which they remain at 1990 levels. As in our earlier work, our focus is on the impacts of a carbon constraint upon the cost of energy. We do not attempt to calculate the benefits of a global carbon agreement.

MODEL OVERVIEW AND KEY ASSUMPTIONS

The analysis employs the Global 2100 model, and analytical framework for exploring the costs of carbon emission limits. We chose the name Global 2100 in order to emphasize both the global nature of the carbon emissions problem and the need for a long-term perspective. For each region, a dynamic nonlinear optimization is employed to simulate either a market or a planned economy. The model is benchmarked against 1990 base year statistics, and the projections cover eleven ten-year time intervals extending from 2000 through 2100.

Three of the model's demand parameters are crucial to the costs of a carbon constraint. The first is the rate of GDP growth. This rate depends on both population and per capita productivity trends. In parallel with the slowing of population growth during the twenty-first century, there will be a diminishing rate of growth of GDP and, hence, a slowdown in the demand for energy. It is assumed that the potential annual GDP growth rate for the U.S. will be 2.5% from 1990 to 2000, and that it will slow down to 1% during the latter half of the 21st century. These rates represent the average of the higher- and lower-growth cases adopted by the IPCC's Working Group III (1990). Because of energy-economy interactions, the potential GDP growth rates do not uniquely determine the realized rates. Energy costs represent just one of the claims on the economy's output. Tighter environmental standards and/or an increase in energy costs will reduce the net amount of output available for meeting current consumption and investment demands. The potential will then exceed the realized GDP.

Energy consumption need not grow at the same rate as the GDP. Over the long run, they may be decoupled. In Global 2100, these possibilities are summarized through two macroeconomic parameters: ESUB (the elasticity of price-induced substitution) and AEEI (autonomous energy efficiency improvements). If there is sufficient time for the adaptation of capital stocks, most analysts would agree that there is a good deal of possible substitutability between the inputs of capital, labor, and energy. The degree of substitutability will affect the economic losses from energy scarcities and price increases. In our model, the ease or difficulty of these trade-offs is summarized by ESUB. The higher is the value of ESUB, the less expensive it is to decouple energy consumption from GDP growth during a period of rising energy prices. When energy costs are a small fraction of total output, ESUB is approximately equal to the absolute value of the price elasticity of demand. In Global 2100, this parameter is measured at the point of secondary energy production: electricity at the busbar, crude oil and synthetic fuels at the refinery gate. The numerical value of ESUB has been taken to be .40 for the U.S.

In addition to the reductions in energy demand induced by rising energy prices, there is also the impact of autonomous energy efficiency improvements. Nonprice efficiency improvements may be brough about by deliberate changes in public policy, such as speed limits for automobiles. Energy consumption may also decline as a result of shifts in the basic economic mix away from manufactured goods and toward more services. Thus, the AEEI summarizes all sources of reductions in the economy-wide energy intensity per unit of output that might occur at constant energy prices. For the present analysis, we adopt an AEEI of .5% annually. This is reasonably consistent with the U.S. experience between 1960 and 1990.

Turning to the supply side of the model, Global 2100 distinguishes between electric and nonelectric energy. Each technology is characterized by its unit costs, fuel requirements, carbon emission coefficients, dates of introduction, and maximum rates of expansion or

TABLE 1. IDENTIFICATION OF ELECTRICITY GENERATION TECHNOLOGIES

Existing Technolgies	Identification
HYDRO	Hydroelectric, geothermal, and other renewables
GAS-R	Remaining initial gas fired
OIL-R	Remaining initial oil fired
COAL-R	Remaining initial coal fired
NUC-R	Remaining initial nuclear

New Technologies	Identification	Earliest Possible Introduction Date[a]	Estimated Cost[b] (mills/kWh)
GAS-N	Advanced combined cycle, gas fired	1995	34.3[c]
COAL-N	New coal fired	1990	51.0
ADV-HC	High-cost carbon free	2010	75.0
ADV-LC	Low-cost carbon free	2020	50.0

a. Estimated year when the technology could provide .1 trillion kwh (approximately 20 GW of installed capacity at 60 percent capacity factor).
b. Based on 1990 dollars.
c. Based on price of gas in 1990. Gas prices are projected to rise over time.

decline. The supply expansion constraints are not rigid upper bounds but soft constraints. Growth may be accelerated but at a rising marginal cost.

Table 1 identifies the alternative sources of electricity supply. The first five technologies represent existing sources: hydroelectric and other renewables, gas-, oil- and coal-fired units, and nuclear power plants. The second group of technologies includes the new electricity generation options that are likely to become available. They differ in terms of their projected costs, carbon emission rates, and dates of introduction. These technologies are intended to be representative of existing and future options within the United States.

Table 2 identifies the nine alternative sources of nonelectric energy within the model. The list is headed by OIL-MX, imports less exports of crude oil. Petroleum is the international "swing" fuel and its price is crucial to any near- or medium-term projections of energy supplies and demands.

All other carbon-based fuels are ranked in ascending order of their cost per gigajoule (GJ) of crude oil equivalent. The last expensive domestic source is CLDU: coal employed for direct uses in industries such as steel and cement. In most scenarios, its maximum potential growth rate is taken to be zero. Next in the merit order are domestic oil and gas. These domestic resources are available at a constant marginal cost but are subject to upper bounds on extraction rates based on a Hotelling-type model of reserves and resource depletion.

TABLE 2. NONELECTRIC ENERGY SUPPLIES

Technology name	Description	Carbon emission coefficient (tons of carbon per GJ of crude oil equivalent)	Unit cost per GJ of crude oil equivalent (1990 Dollars)
OIL-MX	Oil imports minus exports	.0199	4.00 in 1990 rising to 8.40 from 2040 onward
CLDU	Coal - direct uses	.0241	2.00
OIL-LC	Oil - low cost	.0199	2.50
GAS-LC	Natural gas - low cost	.0137	2.75[a]
OIL-HC	Oil - high cost	.0199	6.00
GAS-HC	Natural gas - high cost	.0137	6.25[a]
RNEW	Renewables	.0000	8.20
SYNF	Synthetic fuels	.0400	8.33
NE-BAK	Nonelectric backstop	.0000	16.67

Note: The source of most of these carbon emission and cost coefficients is Energy Modeling Forum 12 (1990).

a. To allow for gas distribution costs, $1.25 per GJ is added to the wellhead price.

Two broad categories of carbon-free alternatives are included in Global 2100: RNEW (low-cost renewables such as ethanol from biomass) and NE-BAK (high-cost backstops such as hydrogen produced through photovoltaics and electrolysis). The key distinction is that RNEW is in limited supply but NE-BAK is available in unlimited quantities at a constant but considerably higher marginal cost. In the absence of a carbon constraint, SYNF (synthetic fuels based on coal or shale oil) would place an upper bound on the future cost of nonelectric energy. For a detailed description of Global 2100 and its underlying data base, see Manne and Richels (1992).

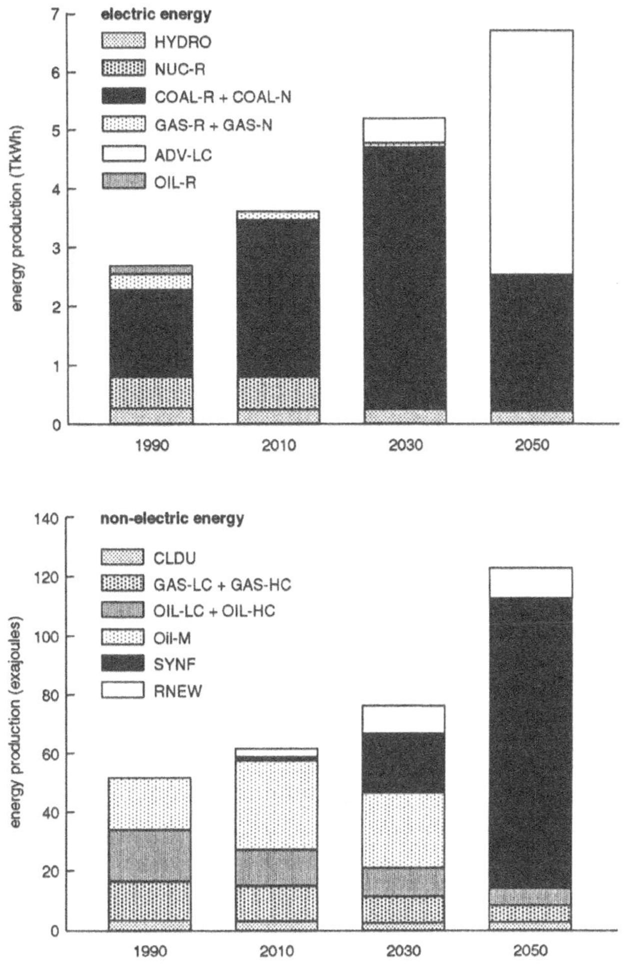

FIGURE 1. U.S. ENERGY USE

A BUSINESS-AS-USUAL ENERGY FUTURE

The ease of difficulty of achieving an emissions reduction target will depend not only on the current composition of the energy sector but also on how the energy sector is likely to evolve in the absence of carbon limits. Would a business-as-usual energy future mean continued heavy dependence on carbon-intensive fuels? Alternatively, would the price mechanism lead toward nuclear power, carbon-free renewables, and/or accelerated energy efficiency?

We begin by projecting the evolution of the energy sector under business-as-usual conditions. Figure 1 presents successive snapshots of the electric and nonelectric sectors at several points in time. 1990 is based on the historical record. The figure highlights the importance of coal to the U.S. electricity industry. On the supply side, there are two principal alternatives to coal: natural gas and AD-LC (an advanced low-cost carbon-free technology). If natural gas prices remain at 1990 levels, gas-fired electricity would represent an attractive option. However, geological resource constraints (and competing demands from the nonelectric sector) are likely to lead to significant gas price increases. Gas-fired electricity is eventually expected to lose its competitive advantage over coal.

The other low-cost alternative to coal is the carbon-free technology, AD-LC. Any of a number of technologies could be included in this category: solar, nuclear, biomass, and others. We assume that at least one of these options will be attractive for economic reasons alone. We also assume that there are constraints on the rate at which it can enter the marketplace. If it were introduced in 2020, it would take on an increasing share of the electric load thereafter. This automatically leads to a substantial reduction in the rate of growth of carbon emissions.

On the nonelectric side, the story is more complicated. Through 2030, oil imports are the largest source of supplies. Since we assume that global oil and gas resources are limited, the exhaustion of these conventional fuels will lead to a run-up in nonelectric energy prices. As crude oil prices approach $50/barrel, two new sources become attractive: synthetic fuels (SYNF) and low-cost carbon-free renewables (RNEW). With RNEW supplies limited to 10 exajoules per year, SYNF would eventually displace imported oil and become the marginal source of nonelectric energy.

A long-term increase in energy prices will affect not only supplies but also demands. Consumers will be motivated to cut back on their consumption of energy. Since ESUB (the elasticity of price-induced substitution) is .4, this means that a 1 percent price increase will lead to a decline of 0.4 percent in the demand for energy.

Figure 2 shows the carbon emissions time path corresponding to our business-as-usual scenario. After 2030, the emissions growth is concentrated in the nonelectric sector. Coal-based synthetic fuels (the long-term marginal source of supply) are more carbon-intensive than those that they have replaced (crude oil and natural gas). Conversely, in the electric sector, coal is eventually replaced with carbon-free technologies. By the end of the twenty-first century, no carbon is released by the generation of electricity.

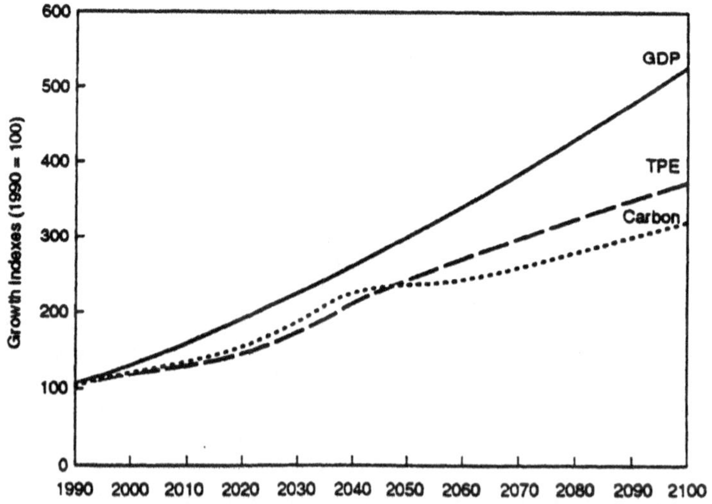

FIGURE 2. CARBON EMISSIONS UNDER BUSINESS AS USUAL

Despite the absolute growth in emissions, there would still be a slowdown in the percentage annual growth rate under business as usual. Between 1950 and 1990, U.S. emissions grew at an average annual rate of 1.8 percent. The growth rate during the twenty-first century is projected to be only 1.1 percent per year. Figure 3 is useful in understanding the reasons for this slowdown. Part of the explanation is a decline in the GDP growth rate to 1.5 percent annually. Although this compounds to a fivefold increase during the twenty-first century, the annual rate is low by historical standards. According to our calculations, carbon emissions and TPE (total primary energy) will grow even more slowly than the GDP, for two reasons: fuel switching and reduction in energy use per unit of output. The contribution of each is indicated by Figure 3.

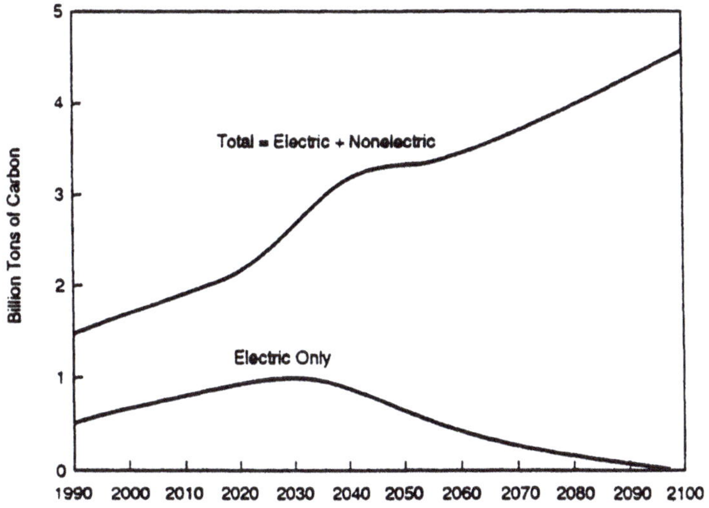

FIGURE 3. GROWTH INDEXES UNDER BUSINESS AS USUAL

IMPACTS OF A CARBON CONSTRAINT

We now estimate the costs of limiting U.S. carbon emissions. There has been enormous debate in recent years over emission targets - primarily for the next decade, but also for the longer term. the focus of the Intergovernmental Negotiating Committee (INC) has been on 2000. The E.C. (and others) are proposing that industrialized countries restrict emissions to 1990 levels by that year. The U.S., on the other hand, is advocating greater flexibility with regard to targets and timetables.

There is even more disagreement over long-term emission goals. Some proposals have aimed only at slowing the rate of growth of emissions; others have set targets of ultimately reducing emissions to half their current levels. For our first set of calculations, we explore a two-stage emissions reduction scenario. Emissions are initially stabilized at 1990 levels by 2000 and then gradually reduced by an additional 20% by 2010. Although these targets are not as stringent as those contained in some proposals, they nevertheless represent a substantial reduction in future emissions when compared with a business-as-usual energy future.

Figure 4 shows the GDP, TPE (total primary energy), and carbon projections corresponding to these limits. Over time, there is a change in the respective roles of conservation and of fuel switching. During the early decades (1990-2020), noncarbon-based supply options are severely limited. Conservation plays the dominant role. Through higher energy prices, GDP and energy growth are virtually decoupled. During the later decades, there are fewer constraints on the rate of market penetration of new sources of energy supply. Once the economy overcomes the introduction limits on carbon-free backstops, there is scope for growth in energy consumption. Fuel switching is then the principal means of adapting to the carbon constraint.

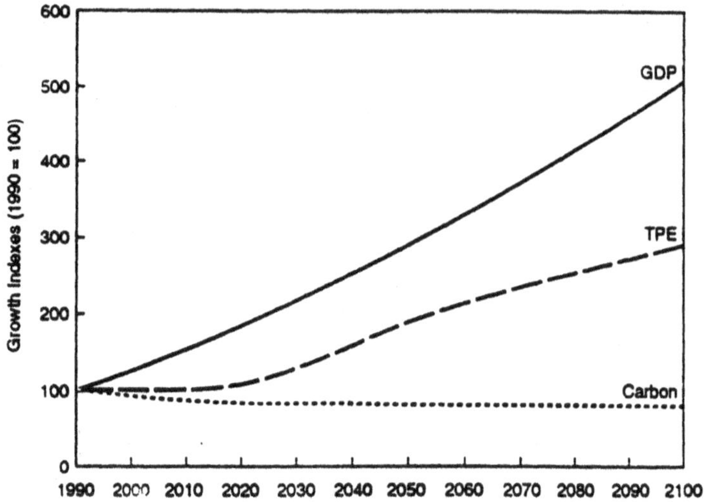

FIGURE 4. GROWTH INDEXES UNDER CARBON EMISSIONS REDUCTION SCENARIO

Using Global 2100, we add together the costs of a carbon constraint throughout the economic system and calculate the annual losses in GDP. Figure 5 shows how these might vary over time. By 2000, the costs approach 1 percent of conventionally measured GDP. At that point, the rise in energy prices begins to have a significant effect on the share of gross output available for consumption and investment. By 2020, the GDP losses exceed 2 percent. Thereafter, this percentage grows to 2.5 percent and then remains constant. Adding over all the years from 1990 through 2100, the present value of the consumption losses would be $1.35 trillions, discounting to 1990 at 5 percent per year.

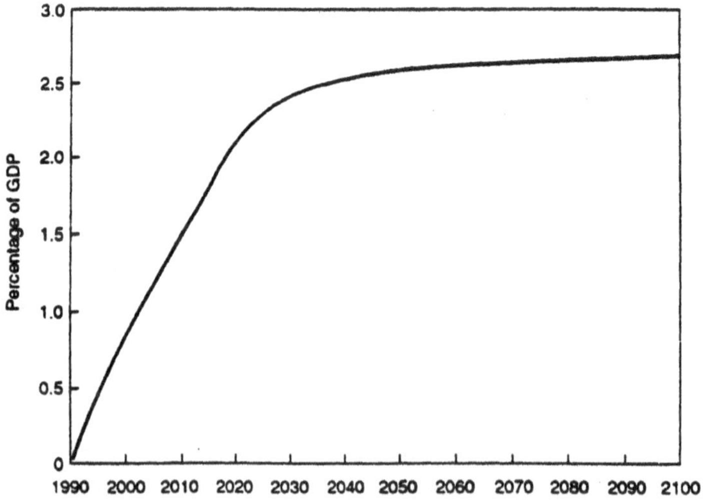

FIGURE 5. ANNUAL LOSSES DUE TO CARBON LIMIT

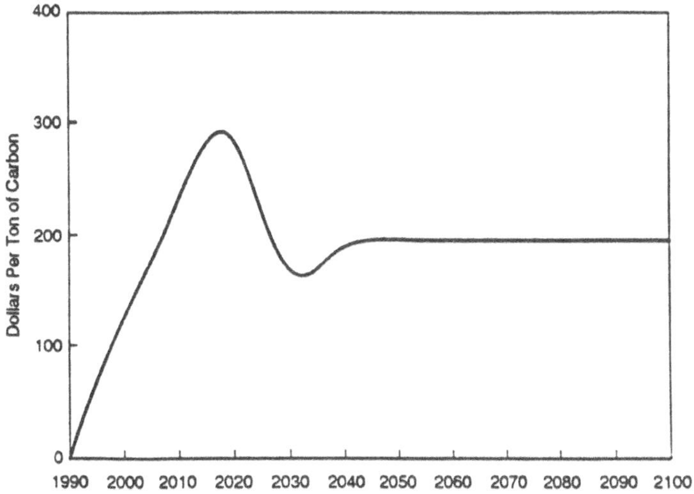

FIGURE 6. CARBON TAX RATES (VALUE OF EMISSION RIGHTS)

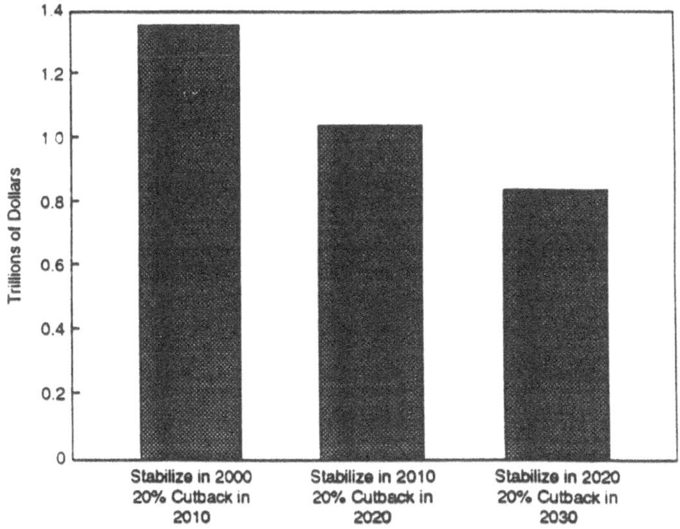

FIGURE 7. EFFECT OF TIMING ON COSTS OF CARBON CONSTRAINT (1990-2100)

TIMING AND THE COSTS OF A CARBON CONSTRAINT

We next look at how the timing of the carbon constraint affects overall costs. To understand this relationship, it is instructive to look at the marginal cost of emissions abatement and how it changes over time. Figure 6 shows the carbon tax path that would be required to achieve the emission targets laid out in the previous section. Note that there are significant variations in the rate of the carbon tax at different times. The reason for the overshoot in the early decades of the 21st century is insufficient supplies of carbon-free alternatives. This is due to constraints on the rate at which new technologies can be introduced into the marketplace. With limited alternatives, consumers are willing to pay a high price to burn carbon-based fuels. The tax must be sufficiently high to discourage these demands.

The spike in the carbon tax path points to several interesting questions. What if the carbon constraint described earlier were delayed by a decade or two to give the economy more time to adapt? What would be the impact on costs? What would be the impact on cumulative emissions? Figure 7 shows how costs vary with the date of the target. Slipping the carbon constraint by ten and twenty years reduces cumulative discounted costs by 25 percent and 40 percent, respectively.

It should come as no surprise that the price tag for a carbon constraint is sensitive to the date of its implementation. The shift to a less carbon intensive economy cannot happen overnight. The time scale for large-scale deployment of new supply technologies is typically measured in decades. Widespread adoption of highly-efficient end-use technologies also takes time. Energy efficient systems are often embedded in long lived durable goods (autos, housing, equipment, structures), and these will not be replaced instantaneously. The process can be accelerated, but at a cost.

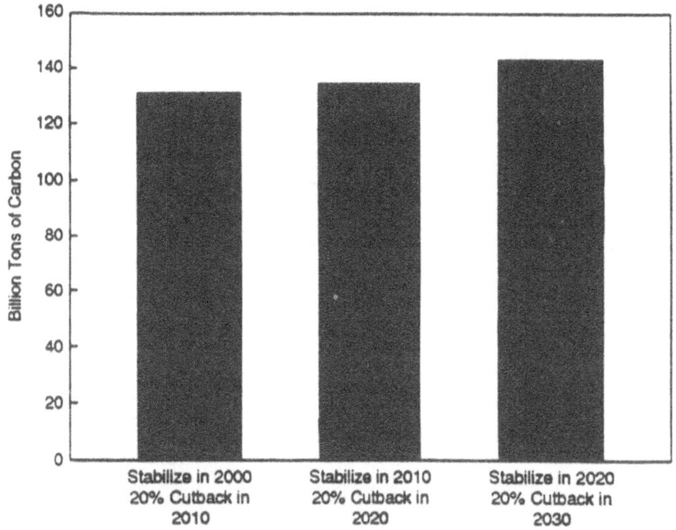

FIGURE 8. EFFECT OF TIMING ON TOTAL CARBON EMISSIONS (1990-2100)

Clearly, flexibility in managing the transition to a less carbon intensive economy is worth a great deal. But what will it cost in terms of incremental environmental damage? This turns out to be the more troublesome question. Large uncertainties remain in our understanding of the greenhouse effect, its likely impacts, and the effectiveness of various countermeasures.

Given the current state of knowledge, it is difficult to assess the costs of a delay. Initial work in this area, however, suggests that the costs may be modest. Schlesinger and Jiang (1991) have looked at the impacts of a 10 year delay in initiating any of the IPCC emission reduction scenarios, and they find that the effect on projected global warming in 2100 is small. Nordhaus (1992) has examined the impact of a 10 year delay in implementing what he identifies as the optimal global policy, and also finds the impact to be small.

It is straightforward to calculate the effect of a U.S. delay upon its cumulative emissions over the next century. From Figure 8, we see that a ten year delay increases total U.S. emissions over the 21st century by between 3 to 4 percent. Delaying implementation by 20 years increases cumulative emission by 9%.

This discussion suggests an additional experiment. Suppose that the focus is on the cumulative total rather than year-by-year emissions. (For an earlier paper where this was suggested, see Blitzer et al., 1992.) After all, carbon dioxide has a residency in the atmosphere that is measured in terms of 50 to 200 years. What if countries were permitted to emit more in the early years as long as they balanced their account by, say, the middle of the next century? With proper anticipation, this could provide much needed flexibility during those years in which the marginal cost of emissions abatement is high. And the "payback" could occur when low-cost technological alternatives are more plentiful.

Figure 9 provides such an example. The solid line refers to our initial set of targets. The dotted line shows a scenario which results in the same cumulative emissions, but the year-by-year path is different. Here emission rights are allocated so as to minimize overall abatement costs. "With borrowing," discounted consumption losses over the 21st century add up to approximately $1 trillion (at a 5% real discount rate)." "Without borrowing" losses are 35% higher: $1.35 trillion. Figure 10 compares carbon taxes for the two scenarios. Note that taxes are positive in both cases, but are far lower in 2000 when the more gradual implementation policy is adopted.

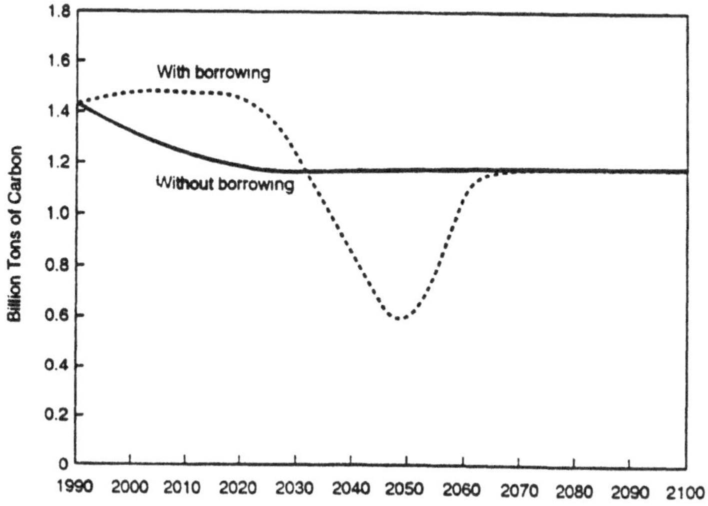

FIGURE 9. CARBON EMISSIONS WITH AND WITHOUT BORROWING

THE YEAR 2000 AND BEYOND

Discussions of the cost of emission controls often focus on a single year. In particular, there has been considerable debate over how much the E.C. proposal would cost the U.S. at the turn of the century. According to Figure 5, costs could approach 1% of GDP by that year. This estimate, however, is quite sensitive to assumptions about longer term goals. Energy investments are often long lived. Consumers and producers will be guided by their expectations for the long term. Figure 11 highlights the importance of the ultimate target. Note that the two scenarios have similar targets for 2000, but differ thereafter. Costs are lower for the second scenario - not only for subsequent years, but also for the year 2000.

In any event, the focus on 2000 may be misplaced. According to Figure 11, costs are apt to be a good deal higher in later years. This is because emissions are expected to grow under business as usual. As a result, the necessary reduction in emissions required to maintain stabilization will also grow. For example, it will be necessary to remove approximately twice as much carbon in 2010 as in 2000 in order to achieve 1990 emission levels. Unless the marginal costs of emissions abatement drops a good deal faster and further than we believe likely, the costs to the economy can only increase over time. Focusing on 2000 may give misleading indications about the overall costs of stabilizing emissions.

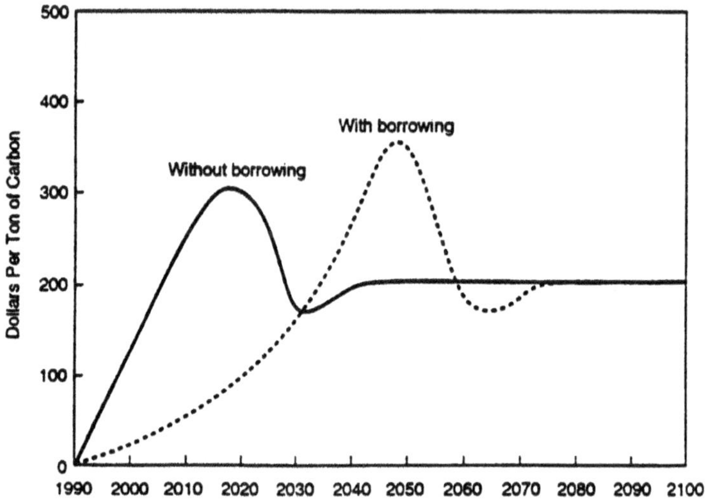

FIGURE 10. CARBON TAXES WITH AND WITHOUT BORROWING

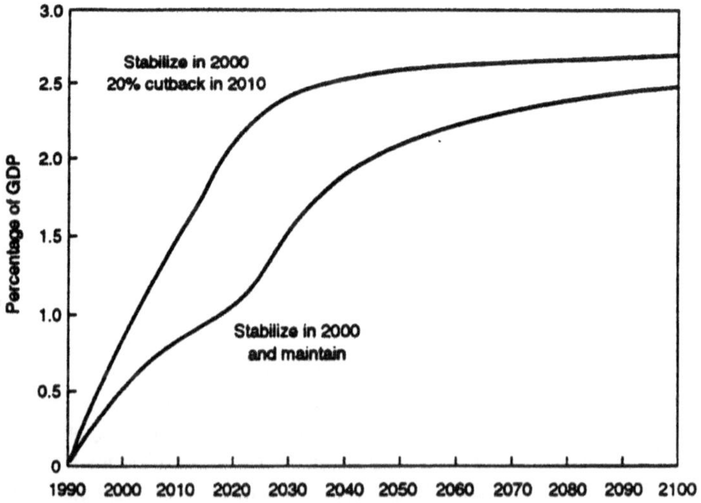

FIGURE 11. ANNUAL LOSSES DUE TO CARBON LIMITS

REFERENCES

Blitzer, C.R., R.S. Eckaus, S. Lahiri and A. Meeraus. 1992. "Growth and Welfare Losses from Carbon Emission Restrictions: A General Equilibrium Analysis for Egypt," MIT Center for Energy Policy Research, February.

Bradley, R.A., E.C. Watts, and E.R. Williams. 1991. "Limiting Net Greenhouse Gas Emissions in the United States," U.S. Department of Energy, Washington, D.C.

Energy Modeling Forum. 1990. "First Round Study Design for EMF 12," Stanford University, December.

Manne, A.S. and R.G. Richels. 1992. Buying Greenhouse Insurance - the Economic Costs of CO_2 Emission Limits, The MIT Press, Cambridge, Mass.

Nordhaus, W.D. 1992. "Rolling the 'Dice:' An Optimal Transition Path for Controlling Greenhouse Gases." Paper presented at the Annual Meeting of the American Society for the Advancement of Science, Chicago, February.

Schlesinger, M.E. and X. Jiang. 1991. "Revised Projection of Future Greenhouse Warming," Nature 350, March.

Working Group III. 1990. "Formulation of Response Strategies." Intergovernmental Panel on Climate Change, Washington, D.C.

THE ALTERNATIVE ENERGY FUTURE

Bruce Henning

American Gas Association

Of late, the natural gas industry has had much of its attention focused on 825 North Capitol Street, the Federal Energy Regulatory Commission, and the restructuring that has undergone in our industry. The competition and efficiency that has been unleashed in the last couple of years to the natural gas industry carry from the creation of the commodity to sale of the service at the burner tip, and more changes are inevitable in all energy sectors.

Those of us who are here in Washington know there's a new troika being discussed in energy futures, that of conservation, renewables and natural gas. A little more than a year ago, Richard Farman from Southern California Gas Company, foresaw these forces melding and viewed that there were certain commonalities between those sectors of the energy industry. We wanted to go forward and create an alliance between those sectors of the energy industry and determine where we could work together.

As a first step in that, the trade associations for those energy industries, the American Gas Association, representing the gas industry, the Alliance to Save Energy, representing the conservation industry, and the Solar Energy Industry Association representing the renewables industry, got together to decide what common visions we had in terms of where the energy market was likely to go. We figured that from the initial stages, we were going to do a whole lot better if we identified, through a formal study process, where we had agreement on what energy markets were likely to look like in the future, and work on policy issues later.

When we got together relatively quickly, we determined that there was one common perception that carried through. That was that most forecasts of energy markets were not reflecting the full importance of market forces and technology development in the analyses which they were performing. This was very important, because it's clear that if you have large movements of primary energy, then there are likely to be significant economic tradeoffs between greenhouse gas emissions versus economic growth. But if, however, primary energy consumption is not likely to grow as much relative to GDP, and the movement from carbon intensive fossil fuels to less carbon intensive fossil fuels is a likely outcome of energy markets, then the economic tradeoffs which are portrayed are going to be significantly less.

When we went forward on this, we constructed a study of an alternative energy future, which was designed to look at the possibility that most economic forecasts are dramatically overstating the cost tradeoffs associated with control of greenhouse gases.

The 1990 energy balance for the United States was just under 84 quads of primary energy consumption. The Department of Energy forecasts primary energy demand increasing to 106 quads. AGA's own baseline projection of business as usual shows about 95 quads of energy consumption in the year 2010, and the newly created Alternative Energy Futures study, which shows relative stability in overall primary energy consumption, winds up in the year 2010 at just under 83 quads.

The forecast between the alternative energy futures study and the AGA terra-based case use identical assumptions in terms of overall economic growth. There are a number of other factors which we'll get into which differentiate the scenario.

When we consider how this balances out in the alternative energy future, we find that we have increased reliance on natural gas. The share of natural gas grows dramatically, from just under 20 quads in the year 1990 to over 27 quads in the year 2010. Oil consumption declines and coal consumption declines rather dramatically. Renewables basically double during this period, and we assume a stabilization of nuclear power which reflects a decline as those units are retired.

When our study team began to discuss these results, the renewables industry thought that we were dramatically underestimating their potential contributions. But we are not looking at a policy induced case, where there were additional incentives, but rather questioning what was economically feasible under a series of assumptions.

Let's look for a minute as to what those kinds of key differences were. The Alternative Energy Future study assumed a very aggressive market penetration rate for more efficient technology. We looked at the growth in technology and the cost estimates associated and advanced their penetration into the market. The AEF also fully incorporated recent regulations for standards of appliance energy efficiency and conservation measures. Included in that are trends towards integrated resource planning and demand side management, not only on the electric utility side, but also on the gas side. As you're all probably aware, the energy policy act includes both Title I and Title III standards for IRP at the state level.

These plans include as their demand side management options not only conservation and load levelling programs, but also fuel substitution programs where those meet a total resource cost standard and result in lower total primary energy demand. Finally, we looked at the removal of market barriers that were inhibiting rational economic energy decision making. We have state implementation plans for the Clean Air Act which eliminate the most economic choices. Often that will include the co-firing of natural gas for reductions in SO2 emissions and NOX emissions as well.

What is really driving these overall differences in the level of energy consumption is the elimination of the state and federal regulatory biases that discourage energy efficiency and encourage carbon intensive fuels. The responses to the DOE notice of inquiry on what are the impediments to natural gas consumption show that there are a number of states, primarily coal-producing states, which notably favor coal in electric generation. Additionally, you have imbalances in the way electric utilities and natural gas utilities can market their respective fuels. Often these things are included in the integrated resource plans but the results, through incorrect analysis, wound up increasing the overall energy consumption rather than decreasing it.

We are also looking at a reallocation of federal R&D money to renewables and energy efficiency as well as to natural gas. Much progress has been made in the last several years in terms of the DOE budget, but there's clearly room for improvement. Continued vigorous expansion on IRPs including fuel substitution programs and implementation of the NACA standards, reflect fully in the model. Quite honestly, we feel that most models do not accurately reflect the regulations on the books today.

What is the result of this different look at what the energy future might hold? Carbon dioxide emissions from these categories, at 1990 levels, were about 5.7 billion tons from fossil fuel. If you compare the DOE forecasts from 1992, there is an increase of up to 7.2 billion tons. Our own projections, which contain a little more gas, show growth to a level of about six and a half billion tons, but under the alternative energy future study, incorporating overall efficiency gains, holding energy consumption fairly level, and decreasing the reliance on carbon intensive fuels, projections show a 12 percent decline to about five billion short tons.

In addition to this, the net loading of the criteria pollutants associated with each of these futures, results in an overall reduction in the aggregates of all the criteria pollutants associated in these scenarios. There are reductions from the baseline, and combined, are over ten million tons.

The question of whether or not you will actually see this overall reduction is problematic. But it is clear that if you change the overall forces associated with the energy markets themselves, you can dramatically reduce the costs of achieving any given level of criteria pollutant emissions.

As soon as we came forward with this particular study, with the debate going on as to whether carbon restraint is likely to increase or decrease economic activity, that reduction came directly to the forefront. We were going through this study procedure just before the Rio conference, so we decided to look at the issues of economics associated with this.

When you look at the available literature, you quickly find that most models, whether they're showing growth in the economy as a result of carbon constraints or whether they're showing declines in the economy as a result of those carbon constraints, are primarily functioning as a result of their fiscal and monetary policy implications to the overall economy and not on what the overall balance of energy markets is looking like. The implications that you see in most of those macro models are related primarily to the economic assumptions and not to the energy market assumptions.

As a result of that, we decided to concentrate on the energy markets directly. We started looking at consumer energy bills, the overall levels of what the consumer is winding up paying under this model. By the year 2000, under those consumer energy bills, there will be a saving of about $68 billion in annual real expenditures. By the year 2010, those savings go to a level of $137 billion in annual consumer expenditures. In each of the cases, when you compare the annual outlays for the additional efficiency measures and development of the domestic U.S. energy industry to the consumer savings, the net result is positive.

In addition, the scenario has significant balance of trade implications. You are displacing foreign oil, you're increasing natural gas slightly, and you save on the order of $42 billion in imported foreign oil.

Finally, the labor implications of this scenario were extremely important. In the overall impacts, there was slightly higher domestic oil and gas exploration. As a result, there were some increases in the overall level of employment in the oil and gas industry, but there was considerably less activity in the coal area, and a decline in the coal labor intensity. But what is not realized by many people is that right now, the labor intensity for oil and gas exploration exceeds that for coal exploration almost by a factor of two. So substituting a quad of natural gas produced domestically for a quad of coal produced domestically, actually increases the number of jobs in the domestic economy.

The final component is that of the energy efficiency industry. Energy efficiency is extremely labor intensive. You're potentially looking at a net increase of employment in the year 2000 of 175,000 jobs.

So when we're looking at an alternative energy future, we're looking at a balancing of energy efficiency penetrations that is achievable through pushing forward and implementing technological improvements. They are dramatic in terms of their outcome.

In the residential energy sector, what kinds of numbers are we looking at changing? During the decade of the 1980s, the average household reduced their energy consumption by about 1.4 percent a year. In our own base case, we are looking at a decline in that rate of energy improvement to a level of about 1.2 percent a year. In the alternative energy future study, we accelerate that to about 1.6 percent a year, but there are no wholesale changes in the rates of penetration of technologies. We are improving the overall resource allocation and using natural gas where it's most efficient for heating, for water and for electric generation, and improving the overall efficiency of that electric generation through the substitution of combined cycle units. Most importantly, we recognize that market forces can play an important role in constraining carbon emissions. This is not a policy induced scenario. This is what can happen through technological progress.

In the years that I've spent in energy modelling, there are three rules that have developed. One is that inevitably, demand is over estimated because the implications of the technologies are not well understood. That's the hardest thing for any one of these models to forecast.

Secondly, on the supply side, the issues of resource depletion are also over estimated. You under estimate the technologies for resource extraction. The combination of those two mean that the overall balance of energy usage is going to be lower and it's going to be cheaper. Reliance on price-induced measures to control carbon are going to be very difficult.

Rather, what you have to rely on is the movement of the market forces and efficient use. Are we on a track for this alternative energy future? Right now, I'm sad to say, no, we're not. We have not reallocated those R&D dollars. We have not removed those state impediments toward the rational choice of pollution control measures. We have not dealt effectively with the issues of fuel substitution in integrated resource planning. Those are the challenges ahead.

In earlier papers there was some discussion about the role of methane. Yes, in fact, methane is a strong greenhouse gas, there's no doubt about that. The exact level of multipliers that one would use to compare with CO_2 is largely dependent not only on some uncertainty in the science, but also on how you aggregate it over time and how you deal with the differential half-lives of methane versus carbon dioxide. But one thing you should be very clear about. In the natural gas industry, when methane is pumped out of the ground and used efficiently, is a very, very small contributor to the overall methane emissions.

EPA, looking at this with the Gas Research Institute for an entire study program, has concluded that it's about one percent loss of throughput, one percent of the total gas produced and then delivered to the end user winds up being emitted. I know there are lots of other numbers that are much larger that are floating around, and in some respects it is a fault of our own reporting of unaccounted gas use.

The industry is going forward to even lower than one percent reduction. It has a voluntary program worked out with EPA called the natural gas star program, where we're exploring cost effective mechanisms for reducing it even further.

So I'm really very pleased to look at the important role that natural gas can play.

THE STATE PERSPECTIVE

Charles Imbrecht

California Energy Commission

In the slightly tarnished, but nevertheless still golden state of California. It's no secret that we've been experiencing a few difficulties, Governor Pete Wilson, has probably experienced virtually every type of social or natural disaster found in recorded history in the first two years of his tenure as our state's chief executive, whether it be disastrous floods in the midst of a drought, incredible holocaust-like fires, social disruption in Los Angeles, not to mention an earthquake or two; finally, the downturn in the national economy and reduction in spending on aerospace and defense industries that are key to the economic strength of California. We've seen it all.

Despite all of those trevails over the last two years, we continue to believe that we have much to be proud of in California. Indeed, we continue to set the pace and represent not only national but international leadership in many technological and environmental pursuits. We are most proud that today, California ranks 48th in per capital energy costs amongst the 50 states. We rank 49th in energy costs per dollar of gross state product and we rank 45th in per capita carbon dioxide emissions in the United States. Indeed, if the rest of the country were to emulate our example with respect to CO2 emissions, the national emission level would be reduced by a full 40 percent. I doubt very much if there would have been much debate about the efficacy of the American role in Rio de Janeiro earlier this year.

Because we recognize that the potential for climate change and all of the various emission characteristics that contribute to climate change are fundamentally functions of how we produce and consume energy in modern society, the legislature in 1988 conferred upon my agency the responsibility of overseeing what was the most comprehensive climate change analysis conducted by any state in the nation. For those of you that are not familiar with the Energy Commission, we are the largest state energy agency in the United States. We have broad responsibilities associated with energy; demand forecasting, supply planning, siting and licensing of major energy facilities in California; responsibility for broad scope research and development activities, ranging from transportation to generation resources, and of course, we also promulgate the most stringent efficiency standards for new buildings and appliances found anywhere in the United States, second only to Sweden in the world.

When we began our evaluation, we did not realize that, as a consequence of the United Nations conference on environment and development, we were conducting an evaluation that was comparable to that which was ultimately recommended in Rio this year. At the earth summit it was concluded under the framework convention that every country in the world, or at least the 155 nations that were signatories to the agreement, should conduct

Global Enery Strategies: Living with Restricted Greenhouse Gas Emissions, Edited by J.C. White, Plenum Press, New York, 1994

an inventory of greenhouse gases and greenhouse gas sinks within their jurisdiction and submit a set of national policies and measures which address climate change to the United Nations on an annual basis.

Indeed, that is what we began to undertake in 1988. In fact, we have produced three separate reports to date in California, reflecting each of the requirements of the international accord. As of March 1, 1992, some 56 countries had made some progress related to inventories for climate change impact assessments. In response to the global climate change convention which was signed in June, about a hundred additional countries will now begin work on climate change country studies.

I thought it might be useful for us to review just briefly our experience with respect to that undertaking in California, because I believe it has some application to other industrialized countries. We think of ourselves as a nation state with the seventh largest economy in the world. Nonetheless, we also recognize that we have an important part to play to provide an example for our sister states.

The three reports that we have produced to date are, first, a report on the potential impacts of climate change to California, ranging from impacts on agriculture, timber and forestry, the level of the sea, the impact on estuaries, wetlands, and our large delta region, and so forth in the state. We have also completed an inventory of the sources of carbon dioxide, methane and chlorofluorocarbons in California. Finally and probably most importantly, we've just recently concluded a report on policy options to reduce the three principle greenhouse gas emissions.

What we have done is now being required of other countries. I think it's important to note that as a consequence of these individual country studies, we clearly will begin to see the framework of the puzzle, about how, on an international basis, we might respond to the concerns of climate change. Obviously, each geographic region and each country, state, et cetera, are going to have different approaches. I might add that that's frequently a function of their own unique circumstances. For example, a small island nation may justifiably focus its country report on the impacts of sea level rise and less on how to mitigate greenhouse gas emissions.

In California, we realize we contribute a great deal to developing global climate change mitigation measures because of our level of policy and technical sophistication with respect to energy and most other environmental matters as well. Today, California contributes about one and a half percent of the total global CO_2 emissions. That's comparable to most of the industrialized European nations. This combination of circumstances as a consequence has caused us to focus most of our effort currently on emission mitigation and less on impact and adaptation policies.

When we undertook our analysis and our inventory, obviously the first question was selection of a base year from which to draw comparisons. In our case, we chose 1988 because that was the year in which we were given the statutory responsibility. Moreover, we were facing a debate during the 1990 elections on a major initiative that had been placed on the ballot by a coalition of environmental organizations which suggested significant 20 and 40 percent reductions in carbon dioxide emissions for the state over a ten and 20 year time frame.

We discovered in conducting our analysis that California, in a very real sense, is not typical of the rest of the United States. To begin with, about 43 percent of total CO2 emissions in our state are directly attributable to the transportation sector as contrasted with only about 15 percent on a national basis. That's obviously a function of our relatively inefficient transportation system. I'll address that more later, but it's obviously an issue on which we have focused a great deal of attention in the decade of the 1980s.

By contrast, we have arguably the cleanest electric generation system anywhere in the United States and perhaps even the world. Our electric generation contributes less than 15 percent of our total CO_2 emissions, as contrasted with the national average, in excess of 30 percent.

So it was clear to us, just from that initial snapshot, that we had to redouble our efforts with respect to the transportation sector. Fortunately, because of a convergence of a variety of public policy requirements, we have been involved in the largest and most aggressive demonstration and development of alternative transportation technologies anywhere in the United States. During the decade of the '80s, California spent more than the rest of the United States, federal and state governments combined, on the development of alternative transportation technologies. That was driven by, the recognition that our transportation sector was a hundred percent dependent upon petroleum-based fuels, and by a desire to build into our transportation system the kind of insurance policy that we enjoy in our electric generation system today, namely, broad diversification of energy sources. That insurance policy, diversification, we believe is essential to insuring stability in prices and rates for the California consumer over an extended period of time. We try to avoid potential additional price shocks caused by changes in production costs, whatever the source of petroleum supply.

Finally, I shall discuss energy security and a reduction of criteria emissions in our efforts to bring our state into compliance with the federal Clean Air Act mandates.

As you are aware, in conjunction with our sister agency, the California Air Resources Board, we adopted about two and a half years ago, the most comprehensive set of tailpipe emission standards anywhere in the world.

I'm sure some of you are familiar with them. We refer to them jokingly as the "LEV Brother" standards. The begin with a transitional low-emission vehicle, and they progress to the introduction of zero emission vehicles beginning in 1998. We are stimulating the development of an electric vehicle industry, not only in California, but elsewhere in the country. As a consequence of the amendments to the federal Clean Air Act adopted two years ago, many other states, particularly in our most populated, urbanized regions, are now in the position to opt up to the California standards. As a consequence, a critical mass of market demand necessary to drive the cost down on new transportation technologies is now well within reach. I can say with some confidence that we are going to see a fundamental change in the nature of transportation, not only in California but elsewhere in the United States over the decade of the '90s. For the first time we will see true commercialization of a broad range of alternatives, each of which represent, when compared to either conventional or reformulated gasoline, significant reductions in CO_2 emissions.

These standards have been adopted and ramp up over the decade of the '90s, culminating in 2003 when at least ten percent of our total market will be met with zero emission vehicles. These standards are expected to improve our emissions levels and bring southern California and other major population areas into closer compliance with the federal Clean Air Act.

We also expect that, as we have seen in the past with respect to our development of renewable energy technologies on the electric generation side, we see 47 major high-tech, aerospace and electronic software firms will form a consortium jointly funded by the state and the federal government to literally turn swords into plowshares. This will begin the re-industrialization of California, converting our capabilities for advanced transportation systems in the air to advanced transportation systems on the ground. We today have over 45,000 people in California employed in the renewable energy industry. We dominate the international marketplace with respect to all of these technologies, and we hopefully will find ourselves in a similar position with respect to alternative transportation technologies as the '90s play out.

I should also mention that in the context of our policy recommendations, we began with the same kind of political climate associated with the climate change debate, as is the case here in Washington, D.C. Obviously, there continue to be significant skeptics with respect to the premise of climate change. We in California have enunciated what we would like to refer to as a no-regrets policy. What we have tried to do initially is to look carefully at all the initiatives we currently have underway, and also to look at other options available to us as a state that are low or zero cost in nature. Most of the measures that we have identified for the state's implementation during the '90s build upon both existing private sector and government activities.

The budget that our Governor proposes in January 1993 reflects significant new initiatives in several areas such as a broad range of research development and demonstration programs associated with clean energy technologies, both stationary and mobile, and improved and more stringent building and appliance efficiency standards. The new California building efficiency standards that take effect January 1993, will result in a full 17 percent reduction in energy consumption in our new buildings, compared to our previous standards which were the most stringent in the United States. Today, California has the lowest per capita growth in energy demand of any state in the country. Quite frankly, that's the only way that we have been able to absorb the enormous population growth that we experienced during the 1980s without a massive capital expenditure construction program.

You read a lot about infrastructure problems in California as a part of the debate over our economic competitiveness. Certainly, back in the 1980s, the energy infrastructure frequently found itself at the top of that list. That is no longer the case. You may hear about water, you may hear about transportation, but energy infrastructure is simply not on the list of initiatives that were necessary to undertake, because we've brought our growth in demand well under control.

During the 1980s, California's population increased by fully 25 percent. For 12 consecutive years, we were absorbing the equivalent of the entire population of the state of Montana on an annual basis. That obviously represents a significant demand upon efforts to provide public services. It is only because of our investment in energy efficiency that we have been able to accommodate it.

We also contemplate much more aggressive information dissemination and a variety of other policy issues. Like the United States and a few individual states, what we've done at this juncture is create a menu of policy options available to the legislature. Hopefully our legislature, with a record number of new members, can overcome some of the partisan struggle we experienced in 1992 in resolving the state's budget issues, and focus on other policy initiatives.

I might add as well that another critical factor is recognizing the importance of estimating anticipated future emission levels, and how we might accommodate additional emission reductions. At the Energy Commission, because of our broad-scale responsibilities to forecast energy demand for all the utilities in our state and for each sector, we have extraordinarily strong analytical capabilities. We collect a great deal of demographic information on population growth and economic development, and as a consequence, we have been able to make what we think are some relatively good projections of the future emissions profile for California.

I should add as well that these calculations were begun in 1988. Because we did not want to go back and re-invent what was already a very complicated process, they reflect conclusions that were reached prior to the adoption of either the California Clean Air Act in 1989 or the amendments to the federal Clean Air Act in 1990. Our projections at this juncture are based upon a 1988 baseline. By the year 2010, CO2 emissions in our state will grow by approximately 24 percent, absent any other initiatives. This represents about a 1.2 percent increase in CO2 per year. However, on a per capita basis, California emissions projections are projected to decline by seven and a half percent over that same time period. Basically, we contemplate, absent no policy initiatives by the state, increasing CO2 emissions, but emissions that are not growing at a rate as fast as our continued population growth.

You might be surprised to learn that we're no longer growing at the rate of one Montana a year, which is about 800,000 net population growth. We're down to somewhere between 550,000 and 600,000 a year, but we still are experiencing very dramatic growth in our population.

At the adoption of the California energy plan, Governor Wilson made a direct statement to the Commission, asked us to go back to the drawing table and evaluate the likely reduction in growth rate in emissions for the state and what kinds of initiatives might be necessary to eliminate the growth altogether or get it down to a flat point on the charts.

We evaluated the feasibility of achieving various CO2 reduction levels. We analyzed not only the effect of existing state environmental and energy regulations and policies which we project were reasonably expected to occur in the future, but also cost effective measures relative to the California CO2 baseline. Obviously, we are hamstrung to a certain extent by virtue of the federal preemption associated with vehicle efficiency in the transportation arena.

Nonetheless, we concluded that existing policies would, in fact reduce the expected rate of increase from the 24 percent down to 15 percent and that is what we already have in place in some of the initiatives. We also concluded that we had additional cost effective options available that would reduce the rate of increase all the way down to six and a half percent, and a very significant 20 percent reduction in per capita CO2 emissions during that same time period.

As I mentioned above, on a per capita basis, California already has an enviable record. In fact, there is no other industrial state that has lower per capita CO2 emissions. Nonetheless, because of the size of our state and the number of people, now over 32 and a half million, we are second only to Texas nationally in terms of total CO2 emissions.

Population increase is obviously a dilemma that we continue to wrestle with in our state. Some suggest that the recession in California perhaps is a blessing in disguise because it's a market driven set of responses to some of those population pressures. Nonetheless, we are not planning for an extended or permanent recession in California, and we're going to make every effort possible to re-invigorate our economy, we recognize that it's important to continue to pursue not only cost effective but also aggressive initiatives that are designed to at least get us down to that six and a half percent growth in actual emissions over the next 20 years, and perhaps get down to a point of no growth whatsoever.

Finally, I would just like to mention that California has long prided itself on being an international leader in the development of prudent and environmentally acceptable energy policies. To date, we have more of each of the renewable energy technologies built and integrated into our electric generation system-wind, geothermal, biomass and solar-than the entire rest of the world combined. As a consequence, there is no energy technology that is operating anywhere in the world today that is not available for inspection within the borders of our state. It was as a consequence of recognition of the vibrancy of this new industry and the large number of people that are now employed in these technological pursuits, as well as the economic contribution these businesses make in terms of tax revenues and economic stimulus to the state, that back in 1986 we initiated the first state international energy export program. Today, we have over 740 companies in California that are participating in the energy technology export program of the Energy Commission. We have memoranda of understanding with 17 foreign governments, and we conduct in the neighborhood of a dozen trade missions where we bring foreign nationals to the state on an annual basis.

The export program was not initiated to deal with political climate change issues, but we consider it now to be an integral part of our responsive efforts. We fully appreciate that if we're going to expect the rest of the world, and particularly the developing world, to follow a more prudent development path than we have in the industrial countries we clearly have to set an example. Moreover, we have to share the benefits of the lessons we've learned in the development of these new technologies, whether they be stationary or mobile in nature.

International success in addressing this issue is going to depend on a sustained effort over an extended period of time. There is no silver bullet, and there clearly are no easy answers or solutions. Providing good examples of win-win public policy that not only lower costs for the consumer, but also mitigate environmental impacts of all sorts is an essential obligation in my judgment not only of the state of California, but of the United States.

I am happy to be a part of this dialogue. I hope that, as a consequence of these and other efforts, we will once again see America taking the leadership role that it most justly requires.

WHY DO THE STRATEGIES DIFFER?

David Montgomery

Charles River Associates

In this paper, I shall try to describe some of the common themes and areas of difference in what has been presented in the preceeding paper, because I think there is a very important set of common themes that emerge. I'll also try to identify what the differences are in approach and conclusion, and some of the reasons for those differences. I hope to help you in assessing what the available evidence is for distinguishing among those approaches and deciding which of them makes more sense. Those reflections sometimes require going a little bit behind the presentations and conclusions to ask where they come from.

In this country we have a policy triangle. The notion of this triangle is that there are three objectives that we have tried to pursue for a long time: performance of the economy and our economic well-being, protection of the environment, and then other energy issues -- energy security and the reduction of oil imports, transition away from insecure energy sources.

The idea of this triangle is that we cannot have as much as we want of all of these. There are tradeoffs in the system, for instance if we want to improve the condition of the environment, we will have to do so at some economic cost.

Likewise, there are some conflicts between energy policy and environmental policy. Promotion of the substitution of coal for imported oil makes sense from an energy security point of view and makes no sense at all from the point of view of improving carbon dioxide emissions.

The other point of view which has been expressed in some of the previous presentations is that we can break out of this triangle. There are no-regret strategies, or the free lunch is indeed possible, and we can have both better economic performance and improved environmental protection at the same time. I would like to try to organize and classify the observations you've heard and then try to assess some of the foundations for them.

Let me begin by trying to summarize briefly what are the key aspects of the previous papers. First, the discussion of the national energy strategy focused on policies, on discussing a specific list of measures and what their implications would be for energy markets, and what their costs would be, estimating the carbon reductions that were likely to come from those policies.

That presentation didn't mention costs. The U.S. approach to the international negotiations on climate policy has been to take the position he articulated as a no-regrets policy. We are taking a set of measures on energy and environmental issues for other reasons. We will treat those costs as being sunk, and simply ask what are the implications

Global Enery Strategies: Living with Restricted Greenhouse Gas Emissions, Edited by J.C. White, Plenum Press, New York, 1994

of those measures for reduction of carbon emissions. The assumption is that the national energy strategy is one whose benefits exceeded its costs, leaving aside any consideration of global environment and therefore, it was recommended to the Congress and adopted. We don't find cost in the national energy strategy, but we do find a fairly consistent discussion of tradeoffs. If we were to go beyond the point where the national energy strategy takes us, which still leaves us with growing carbon emissions over the next two decades, any reduction in those emissions would be costly and we have to assess what those costs are worth.

The presentation from the Environmental Protection Agency also had a policy focus. It talked about specific measures, estimated their carbon reductions, and brought in the concept of cost. We find in their thinking some free-lunch measures. Volunteerism is something for which the benefits presumably exceed the cost. Reductions in agriculture subsidies are measures which would bring about economic improvements to improve the efficiency of that sector, and would at the same time reduce some activities such as growing rice in subsidized water in California or encouraging the production of dairy cattle through milk price subsidies, both of which produce methane emissions. We have some reductions in carbon emissions that are a free lunch.

Other measures were recognized as being costly. Regulatory interventions like CAFE standards or tightening efficiency standards are costs which have to be balanced against the potential gain.

Mr. Richels' presentation on carbon taxes and carbon emissions raises a number of interesting issues. Let me point out that the central conclusion is that carbon taxes reduce GDP by about two percent for a carbon tax sufficient to achieve a specific stabilization goal. There is no free lunch in this sense, carbon emissions reductions are not free.

The Alternative Energy Future takes a very different tack with all of these questions. The presentation began by emphasizing the importance of market forces and technology development, discussed a forecast which contains lower energy use and lower emissions than the AGA terra baseline, and somehow had a claim which said, there are no policies in our case; all we wanted to do was reduce market barriers.

It seems to me that that is, in fact, a policy recommendation, with points like reforming regulation, removing subsidies, and redirecting R&D; but the implication drawn in this study was that these changes would benefit consumers. Consumer expenditures for energy would fall by more than the investments consumers had to make to reduce energy use, and would have other large economic benefits-improving the trade deficit and improving employment.

I have the advantage of some abstracts of these presentations, so I was a little bit ahead of them on those. Characterizing Mr. Imbrecht's presentation, he discussed policies, and in particular policies in the transportation sector, what their implications for emission reductions would be. He then discussed a number of the economic benefits and the idea of trying to find no-regret strategies where we could avoid having to pay additional costs in the reduction of carbon emissions.

There are differences among the studies and the way they were put together. First, we ought to establish in a lively debate on local climate change policy the different weights on the different objectives. That is, some may feel that reducing carbon emissions or the importance of the potential dangers to global climate justify more aggressive measures. Others feel that the costs are not worth bearing and the economy needs stimulation. This is especially important in the global context; when we think of tradeoffs that developing countries may face between investing in development (which brings with it increasing energy use), putting the resources into dealing with hunger and poverty or putting resources into adopting more expensive but more environmentally benign ways of using energy. We might think that there are different weights on the objectives, and that they would lead to substantial differences.

That doesn't seem to be an important part of what we hear today, in talking about climate policy. There can be different assessments of how specific actions can contribute to objectives. Here I think it's important to mention that these are not settled questions in the analysis, and they're going to come up in awkward and thorny ways to make it difficult to judge between the conclusions of different studies. An example is whether and by how much alternative fuels for vehicles reduce carbon emissions, and which fuels can reduce carbon emissions over the entire fuel cycle. Another is performance of demand side management programs as part of the electric utility reform, which is often urged as a no-regret strategy for reducing carbon emissions.

Both the economic side of those demand side management programs and their effectiveness in actual use is subject to very different interpretations and assumptions among different analysts. This is not a difference in perspective on how to think about policy or economics; it's a difference in facts, on how specific measures may work.

I think the most important difference to try to understand, though, is different views on whether there is a tradeoff among objectives or not. There's another variant on this, which is different views on whether to give environmental and energy policies credit for solving what are essentially unrelated economic problems; such as the trade deficit, the heavy deadweight losses from taxation of business income, and other economic problems. These have appeared in the studies as well.

In terms of how they were conducted and presented, these studies differ in three ways. They differ in their outlook for energy markets, although that's not particularly important, whether the studies show rapid growing or slowly growing or stable carbon emissions is not particularly relevant to policy analysis. They differ in how they characterize the costs and benefits of reducing energy use. This is terribly important, whether a study believes that increasing CAFE standards for automobiles will benefit consumers at the same time that it reduces their fuel use or will cost consumers at the same time it reduces their fuel use. Then there is difference in policy recommendations, as I've described. But the place that this matters most is how those policy recommendations are tied to the cost of achieving reductions in emissions.

As far as the outlooks for energy markets go, we've seen different assumptions about energy efficiency and different assumptions of new technologies in the studies. The contrast is probably most clear between the alternative energy future and the work that Mr. Richels presented. These are very different views of what the possibilities are for improvements in energy efficiency and new technologies.

Across these models, different assumptions are made for the baseline, different assumptions are made about energy efficiency and about the cost of new technology. These do not change the answer about whether emissions reductions below the baseline will be costly; they simply set where the baseline is. What they do affect is the cost of meeting targets. So if we were to say we wanted to establish a target of constant emissions, it matters whether our baseline projection is such that we're going to meet that target without doing anything else, or not.

The importance of these assumptions depends on the policy that is under discussion. If the policy is one of targets, then it matters what your baseline assumptions are, because you may be able to sneak under the target without doing anything else. If the policy we're talking about is efficiency standards or carbon taxes of $100 per ton, it hardly matters what these assumptions are in the assessment of whether those measures are costly or not.

The important issue in comparison of these different views is how they approach the costs and benefits of reducing energy consumption. The market view is, tradeoffs are unavoidable in putting together an energy and environmental policy. The benefits of environmental improvement have to be balanced against the costs. I would put into this category the work that Mr. Richels has done, the DOE thinking on the implications of further reductions in emissions beyond those implied by the national energy strategy. Mr. Christopher says I can put EPA in this category. The fact is, EPA has found some measures which offer cost savings and some which do not.

Then there's the free-lunch, and for purposes of disclosure, I found myself firmly in this camp as well. The free lunch view has a number of different ways of being expressed, but essentially it comes to the conclusion that reducing energy use will reduce emissions, and will save consumers money. They go on to argue that it will create jobs, and help the trade balance. In any event, it avoids the need for making hard choices about what should be done. Let me try to illustrate where some of this comes from. The market view says there's a limit to how much we can do. There is a maximum we can get in terms of the economy or environmental protection. We have invested a good bit in improving environmental quality over the past two decades. We could have greater protection of the environment from lower emissions of various kinds, but it would cost us some. It may not be worth trying to meet the air quality standards in Los Angeles; its' going to cost us so much to get that last parts per million out of the atmosphere that it's really not worth it. We should back off a little bit on our environmental regulation and have some more jobs. But we can't do both.

The free lunch view says that we are not doing things as efficiently as we could, and if we fixed them up, there is room for great progress. We could move to the point of greater environmental protection and do a little better for the economy. We could do a lot better for the economy without harming the environment at all, if we just adopted the right kinds of measures.

Why is it that we would not find people making decisions on their own that get them as far ahead as possible in this tradeoff between the environment and the economy? The notion here is that consumers will judge for themselves what it's going to cost to install greater insulation in their attics, or how much more they can pay for a car that has higher fuel economy. They can look at the price of energy they save. They can do some simple calculations of how much will be saved over the life of their investment. They can look at the cost of the investment and make their own decisions.

Why are consumers systematically making the wrong decision? Well, some claim that consumers are ignorant and irrational. Engineering says that the consumer should get a two-year payback, if they're not doing it, we ought to tell them to do it. Most economists and policy analysts prefer to look deeper into it and ask, "is there something wrong with the market? Are there ways in which we can correct what's happening in the market?" These are the usual candidates: market imperfections, specific energy markets, regulatory distortions, subsidies to possible fuels which lead to the use of coal rather than natural gas.

One could assume that the market will choose the right amount of environmental protection. It clearly won't. That's the role of government, to decide how far we want to go in protecting the environment. Individual market decisions won't, public policy needs to. The question is where is out reference point?

There are other economic problems that we could fix by means of environmental policies. That's where the argument about tax recycling comes in. We're not saying that reducing emissions have no cost. We're saying that one way we can cover that cost is by using the revenues for a good purpose. Business taxes are agreed by most economists to reduce investment and to cause serious loss to the economy. If we could find other taxes to substitute for those taxes, we could improve the performance of the economy. Carbon taxes are a candidate, so are other taxes. The difficult issue is that carbon taxes are not the best tax to use for tax reform. The general conclusion of tax councils is to choose a value-added tax or a national sales tax or some other broad tax which does not cause people to reduce one particular form of consumption, but rather applies broadly to all kinds of consumption. This is the cheapest way of financing tax reform. Also, we could argue that a large carbon tax may well cause losses that are greater than the business tax that they would be replacing. The issue here is one of how to put the package together, not whether reducing emissions itself is something that is costly.

One reason why we could find ourselves in trouble is market imperfections. Those are cited in many different studies. Subsidies, regulatory distortions and other imperfections are the ones that most commonly come up. I am not comfortable with the evidence that these market imperfections exist on a broad scale. Subsidies matter only if they affect behavior. The most frequently mentioned subsidies are the tax credits and tax exemptions provided for the production of fossil fuels, and the depletion allowance for oil and gas. These subsidies, whatever other arguments there may be about them, in fact are quite limited in their effect. They only apply to small producers up to a specific amount of production. It's unlikely that they would increase production of oil and gas in the United States much from what it would have been, and they certainly don't increase the consumption of oil and gas, because they don't affect the price at which oil and gas is offered in the market. That's determined by the price of imports. These subsidies are essentially a sump cost; they don't affect decisions on the margin. Much the same could be said of many of the other subsidies that are normally cited in defense of the argument that there are large distortions in energy markets. And we have subsidies in our programs with alternative fuels and conservation as well.

Regulatory distortions are ones which certainly need to be addressed. It's not clear nationwide what the bias is in regulatory distortions. The largest one that's cited in many studies is pricing of electricity. Nationwide, it is not clear that there is much difference between the cost of adding new generation capacity and the price which consumers are being charged for the existing capacity. We've had a lot of very expensive power plants come into the rate base and a fair amount of excess capacity in a lot of places, these have pushed electricity prices up beyond expected levels.

The result is that there are individual cases in some states, where electricity prices are lower than the cost of new generation. It's not clear that there is a systematic bias there.

Let me try to sum up where some of these issues take us. What they suggest is that we may have options where we can improve environmental performance without economic sacrifice. It's important to focus on the types of policies that are recommended, rather than on the general notion that there are policies. We can investigate regulatory reform, investigate situations in which the specific market imperfections might occur, and advise policies that would address those market imperfections.

Taking all of this into account, I would suggest that a way of trying to assess the different studies and perceptions that have been addressed today is to determine what their implications are for whether there is or is not a tradeoff between the economy and the environment. In one case there are immense opportunities for reducing emissions at low cost. In another, the only way to reduce emissions is by sacrificing something in economic terms, and the decision has to be made whether it's worth it.

We may be in a position where some degree of reduction in carbon emissions could be achieved through specific kinds of economic reform. Beyond that level, any further reductions or any efforts to stick with targets which would last for a very long time require making hard choices. I think that the most important question in addressing all this is to what extent are these choices laid out clearly, and then to apply the notions of how much it's worth.

DISCUSSION

MR. IMBRECHT: With all due respect, I find the term free lunch to be a bit pejorative in nature, particularly when contrasted with a market view, and suggesting that somehow we have not applied sound market principles to the evolution of policy in California. I would contend by contrast that we in fact have a much more market driven system in California that was fundamentally changed because we pioneered the entire concept of least-cost utility regulation and all the necessary foundation steps required to draw legitimate and rational choices amongst the various market options available.

In addition, I would note that when you talk about market imperfections, the subsidies that were mentioned in my judgment are the tip of the iceberg. The fundamental subsidy is the subsidy that is ongoing, and that is the security of supply, the fact that our nation finances two battle groups on a year-round basis in the Middle East. I don't believe that's entirely to instill harmonious relations between the Israelis and the Arabs. Obviously, there are other strategic considerations for our country and there is certainly some cost, attributable to security of supply of petroleum that ought to be taken in consideration when you're evaluating whether you are indeed affecting price to the point where you're affecting decision at the margin.

Also, in reference to regulatory distortions. For example, the principal regulatory distortion that we focus on is the dis-incentive associated with the rate structure and the historical methods of rate setting between new supply versus demand side investment. We are the first and still the only state that has begun a small fledgling step to provide a similar incentive to the utility for investment in Demand Side Managment as is available for investment in capital expenditures and for generation and supply.

Those are the regulatory distortions in my judgment that drive decisions, not necessarily to the least cost decision, or the decision that is most likely to be as environmentally benign as possible.

Just to repeat three statistics I used during my remarks. These are all provide by the energy administration division of the DOE, and I will argue any of the sub-points if you care to. But I think it's hard to refute the following. As I said, we're 48th in per capita energy cost, 49th in energy cost per dollar of state gross product. At the same time, we're 45th in per capita of CO2. That suggests to me that it is possible to drive a system down in terms of relative expense and at the same time accommodate reasonable environmental objectives.

MR. CRISTOFARO: I think there may be some difference in opinion here with respect to the labels on the efficiency curve that was presented by Dr. Montgomery. I think that one can believe that there are tradeoffs, but still believe that there are free lunch possibilities, which is probably the camp that I would be in. I don't believe that we're on the efficiency frontier between environmental and economic tradeoffs. I do believe, though, that there are policies under debate that do involve tradeoffs.

MR. RICHELS: I just have one comment. People have a mis-impression that the economic models assume necessarily that the markets are perfectly balanced and there are no inefficiencies in the system. That is not the case. It is very easy to represent free lunch scenarios in the current generation of energy policy models. It's very easy to assume that the current configuration of the economy is optimized based on an incorrect estimation of what the real price of energy is, so these models can be used to include free lunches and reduced cost scenarios for reducing emissions.

MR. BERNABO: Question to Mr. Imbrecht, In your figures, are you accounting for emissions of electricity that California imports? And how much of your achievements come from the fact that you don't have harsh winters; there isn't the forcing function of harsh winters on emissions.

MR. IMBRECHT: Frankly, I expected that benign climate allegation. I've heard it once or twice in the past.

Let me just note for you that we have 16 climate zones in California. There are only 18 climate zones found in the contiguous 48 states. We have everything from extremely cold temperatures to obviously very hot temperatures. But the reason that I don't believe that's an adequate explanation is because the predominant growth factors in California now and for the last decade have been in some of our most extreme climate zone regions, where we have the highest temperature variation on an annual basis and which contribute very markedly to our peak demand period, the summer peak associated with air conditioning. The inland empire, the San Joaquim Valley and the various points east of Los Angeles out in the high desert are where all of the growth is occurring. There is little opportunity for additional population growth in the existing urbanized areas, because there is no land left for development as a practical matter.

Importation costs and emission are in our calculations. Let me just note for you that we currently import somewhere between 11 and 12 percent of our total electricity. On the West Coast, we have also pioneered what we refer to as seasonal exchanges. We avoid the necessity to build capacity either in the Pacific Northwest or in California, because of the fact that they have a winter electric peak and we have a summer electric peak. What most people don't realize is that the northwest Pacific inter-tie sends electricity north about 40 percent of the year and brings electricity south about 60 percent of the year. Of that 11 and 12 percent of our total capacity requirements that are imported, the bulk is the spring hydro that's available from the Pacific Northwest, where the Bonneville system as well as the British Columbia hydro system have substantial over capacity.

Our total mix including all imports and taking worst-case scenarios, we have a little less than four percent of coal based generation. That's primarily the power plants in Arizona and New Mexico that we do buy small quantities from. We buy a little bit from Montana as well.

MR. CRISTOFARO: There are a few things that Mr. Montgomery said that I wanted to respond to. The first is whether or not it's fair to take credit for fixing the problems with tax, as we put forward the notion of a carbon tax. What he said was, if you're going to do some analysis of the fixing of taxes, then you should consider the full range of tax alternatives, specifically that if you were to compare a carbon tax to a value-added tax in looking at deadweight loss, the value-added tax would clearly win.

My response is that that's a possibility. But also, you have to take into account that a carbon tax can be viewed as a revenue raiser and it can also be viewed in the sense that you are internalizing an externality. You avoid damage by imposing the tax. It could be that a carbon tax on purely economic grounds would produce less deadweight loss than the value-added tax, depending on the amount of damage that is avoided.

MR. MONTGOMERY: I agree with Dr. Cristofaro completely on the analytical point. I think it probably serves to sharpen the importance of looking at the benefits of doing something about climate change as well as the cost. What we normally mean when we talk about no-regret strategies is, let's assume there are no benefits from climate change and see what kind of policies would be justified on that assumption, and then if some of those policies also bring lower emissions, well, that's fine.

What Dr. Cristofaro is saying is, we get to a point where we really have to ask, how much are those reductions in emissions worth? If we could do that and put that into the calculation, then we could argue that the carbon taxes can further benefit. But unless we bring those benefits of reducing climate change into the calculation, we can't do it.

MR. CRISTOFARO: But the second point with regard to tax reform is, what is the feasible side of reform? The people who discuss tax reform, discuss the tradeoffs involve not only economic performance but also an equity performance. So the question would be, when you consider all of the factors in the tax game, in addition to the environmental damages that may be mitigated through part of the tax, how does it stack up against other kinds of tax programs, like the value-added tax?

There's no clear answer to any of these questions. I only put forward the notion that it's a complicated problem. At EPA, when we put these ideas forward, we're not really trying to snooker the American public or say that this is the clear unambiguous answer, but it certainly is worthy of public attention and debate in the context of tax reform discussions generally.

MR. RICHELS: One of the questions was whether or not the costs of a carbon constraint is sensitive to the absolute level of the constraint. The answer is certainly.

I should also point out that the issue is a bit more subtle than that. Leading up to Rio, there was an awful lot of discussion about the costs of a stabilization scenario for the United States, stabilizing emissions at current levels. The estimate for the year 2000 is not only sensitive to the level of the constraints for the year 2000, it's also extremely sensitive to the assumptions about the level of the constraint after the year 2000, because most of these models are optimization models, where energy producers and consumers are optimizing the configuration of the energy sector, not only based upon their expectations of what the constraints will be today, but on what the constraints will be tomorrow.

Most of the estimates of the cost of stabilizing emissions by the year 2000 and keeping them stable thereafter were of the order of one to two percent of GDP. If one were to assume though that not only were we to stabilize emissions, but that eventually we were to move to stabilizing concentrations, the ultimate goal of the framework convention, then the cost of stabilizing emissions in the year 2000 are a lot higher, because you're building an energy system very quickly that's devoid of CO_2.

MR. IMBRECHT: I am asked to discuss the California Energy Commission's proposal to apply environmental externality costs to electricity imported into the state for consumption by its citizens. Further, do I see this as a trend that will be followed by other states?

First off, it's not a proposal, it is something that has been accomplished twice now. We are about to adopt the new electricity report for California which is our massive document that includes demand forecasting and supply planning for all of the utilities. I should mention that a state statute was passed about three years ago requiring both us and the Public Utilities Commission to do this, so it's not a discretionary initiative on our part.

This information is used for two separate purposes. One is to operate our production cost models that allow us to freely substitute any generation source for any other generation source in order to try to match most carefully our demand requirements on a seasonal basis, and keep our costs at the lowest possible state. We are not using these costs to outlaw the consumption of electricity or price it out of range from those existing facilities in California.

The second purpose for the externalities is in our resource acquisition decision process, what new resources should we acquire? In that case, obviously the externalities do have some impact on the overall cost effectiveness of the plethora of options that we consider on a biennial basis in the state.

Lastly, do I expect the trend to be followed by other states? The answer is yes. In fact, other states particularly in the West have already been doing much the same kind of work that we have in California. Nevada in particular actually beat us to the punch.

MR. BERNABO: Next, there's a question here that is generally to the panel. It's about Dr. Nordhaus's study that was in Science in November of 1992, which I'm sure the panel members are familiar with, asking whether the optimal tax strategy is five dollars a ton for carbon. Would you comment on that, and to what extent do you feel his strategy is effective in balancing the costs and benefits of CO2 reduction?

MR. CRISTOFARO: I don't think that there is any way of saying whether five dollars is the optimal tax to eliminate the damage of potential global climate change. I think that we have at this point gone beyond asking what an optimal tax should be and then looking at the resulting emissions. We did sign a framework convention which committed us to at least aim at returning to our 1990 levels by the end of the decade. We at least have to evaluate our policies if we're going to be in compliance with this international treaty, which we did sign.

I should add that it's not a mandatory requirement in the sense that you must return to those levels, but we are supposed to design these plans and seriously take a look at whether we can achieve these levels. If we can't, then we are opting out because of economic conditions and other factors. So I think we'll be going beyond discussions of optimal taxes and looking at more specific targets in the future.

MR. MONTGOMERY: I think Nordhaus' work for a long time has been very useful and thought-provoking, because he has focused on the issue of what we know about the consequences of climate change, how that will affect economic activity and therefore people. I think he emphasized two other things which are quite important, first the tremendous uncertainty about all of those linkages, how little we know that would enable us to narrow the bounds on what these costs are, and second, the value of getting more information about those scientific aspects.

Given all that, I think his efforts to come up with a ball park notion of how economies of different kinds are affected by climate change are very useful. The problem is not so much in the economics, because there are lots of economic techniques that we use in many ways in trying to evaluate environmental damages; it's all the scientific steps that it takes to get you to the point where an economist can go to work, and the lack of information about all the things that really matter for understanding climate impacts.

Given all that on the benefits side, which says that it will be extremely difficult to come up with something even within a factor of a hundred of what the optimal tax should be, Nordhaus did a really good job of centering attention on the important issues. I think the real important thing is one that Mr. Richels ought to talk about, the notion of starting with a small tax which esculates on a schedule, and is tied to your notions of what the cost of technologies in the future are going to be. Despite what we're committed to in stabilization in the year 2000, the question is whether a much more gradual approach which gives time for the technologies and which also has implications for the carbon tax is the right thing to look at.

MR. RICHELS: I think Bill Nordhaus puts forward an extremely useful framework for thinking about this issue, regardless of what you feel about the actual numbers. It comes down to balancing marginal costs and marginal benefits. Until we get a better understanding of what the damage function looks like and how the damage function shifts as a function of policy, it's going to be very difficult to say what constitutes sensible policy in this area.

The one thing that I do find that's very interesting from the Nordhaus analysis is that, given his preliminary estimate of what the damage is, what you're really talking about is much smaller emissions reductions from the baseline than you would see from a scenario which either stabilizes emissions at current levels or stabilizes concentrations. Incidentally, for those of you who aren't familiar with the numbers, stabilizing concentrations means reducing emissions by 70 percent below current levels.

GLOBAL ENERGY STRATEGIES: THE PERSPECTIVES OF DEVELOPING COUNTRIES AND ECONOMIES IN TRANSITION - INTRODUCTION

Deborah Bleviss

International Institute for Energy Conservation

As we all know, the issue of global warming and all global energy impacts goes well beyond the borders of the United States. We are looking at two groups which have much in common and also many differences. They have been categorized as the developing countries and the developed countries, although many of us would say the terminology isn't quite correct on either.

The developing countries will be the subject of the first set of papers, and then the developed countries will follow. I want to present some of the relative perspectives of both countries and discuss the need to look at the needs of those countries with respect to global energy strategies.

The first set, which has been termed developing countries, basically are the non-OECD countries that are at various levels of industrialization, for whom fairly rapid economic growth is projected in order to attain of the style of living that we enjoy in the U.S., Europe and Japan. We will have presentations representing three geographic regions. One is from what is now known as Central and Eastern Europe, specifically looking at Russia. Then we will look at India and Asia, and finally we will look at South America. I want to note that Africa is missing from this list. I'm sure that this is because it is not expected to see any rapid economic growth in the near future, or much growth in energy, and hence will not be a major contributor to problems associated with global energy use. I would like to point out that many of the issues that we're talking about today are just as relevant to Africa. I believe Dr. Sturm, in a later paper here on behalf of Latin America, will allude somewhat to the applicability of many of the approaches discussed here to that continent.

I might note that when one looks at the so-called developing countries, you cannot divorce from them the desire to attain better quality of life for all their inhabitants. There are a series of developmental issues which must be addressed, from per capita income to such issues as the division between rich and poor. These things must be held in mind when you consider a variety of energy options for these countries; they cannot be divorced.

The tendency then is to look for ways of providing energy services at the least cost. In many cases providing energy services in situations where there is not already an extensive electricity grid means that renewables and dispersed energy generation have a lot of potential. Not that they don't in the developed world, but there is a terrific potential in these markets, particularly when you underline the least cost.

Also critical is energy efficiency. A lot of infrastructure is being created in these countries, and they can either go down the path of having very inefficient infrastructure which must be fixed later, or we can build that efficient infrastructure in from the beginning. I invite all of our presenters to address those issues.

RUSSIA/EASTERN EUROPE

Alexander Kalinin

Monterey Institute of International Studies

I would like to start with the famous question, "To be or not to be, that is the question. Whether it's nobler in the mind to suffer the slings and arrows of outrageous fortune, or to take arms against a sea of troubles and by opposing, end them." So when I face a sea of troubles whose name is global warming, I would like to analyze first what we know and what we don't know about global warming effects.

Climatic abnormalities of the 1980s, a recent spate of record-breaking droughts, heat spells and hurricanes have led some scientists, the majority of the scientific community, to conclude that global warming is already underway. The five warmest years in the last century have all occurred since 1980, and 1988 was the warmest year on record.

For the last century, the average temperature at sea level has increased by approximately 0.5 Centigrade. It could not be ruled out that we are witnessing the highest rate of global temperature change for the last million years. Sea level has raised by ten to 15 centimeters due to the expansion of seawater and melting of polar ice caps. That's experimental observation.

As a matter of fact, global climate is subject to natural variability. Climatic reconstructions imply the existence of geologic periods with the average global temperature fluctuating widely. It deviated significantly by a few degrees against its contemporary value. The lack of data on global climate variations over hundreds and thousands of years provides insufficient grounds to state that the signal of global warming can surely be detected on the background of natural fluctuations.

This global warming, will be the first time in geologic history when climate change is caused by an artificial, humanity-induced greenhouse effect. The physical essence of the greenhouse effect is well known. If there were no greenhouse effect, the average surface global temperature would be approximately 3.2 to 3.5° Centigrade below its actual value of 14° Centigrade.

What is the impact of the energy system of Eastern Europe and of the former Soviet Union, now the new independent countries of the former Soviet Union, on the greenhouse effect? First of all, I would say that despite all the changes in the political and economic situation in the Eastern Europe and the former USSR the joint electricity system and the joint gas pipeline system of the former USSR are still alive and still functioning.

So let me first tell you that the energy systems of Eastern Europe have extremely high energy intensity. On average, the countries of the Eastern Europe have one and a half as

Global Enery Strategies: Living with Restricted Greenhouse Gas Emissions, Edited by J.C. White, Plenum Press, New York, 1994

much energy intensity in comparison with the economies of the countries of Western Europe and United States. Eastern Europe and the former USSR explain their relatively higher contribution to greenhouse gas emissions because they have the common feature of energy derived mostly from coal and shales.

The Soviet Union and the former Soviet republics, are different, because gas supplies comprise about 40 percent of the total energy balance. This is beneficial in the context of the global warming problem, because the amount of greenhouse gases emission is lower.

Using those figures, it's not difficult to compare the equivalent greenhouse emission of Eastern Europe and the former USSR. I draw your attention to the fact that at least 10 to 20 percent of the greenhouse emissions is due to the contribution of other than CO_2, including methane and nitrous oxide. But both methane and nitrous oxide are produced by the energy systems, and we should take into account their contribution to the equivalent greenhouse emission.

There are four sources for energy used in Eastern Europe; brown coal, hard coal, petroleum and natural gas. These are arranged in order of their contribution to greenhouse gases.

Two countries in this area are extremely dirty countries in the sense of greenhouse emission. These are Czechoslovakia and Poland, both with a high share of consumption of coals in their energy balance.

Despite that the former USSR had a very high energy intensity and very high volumes of primary energy consumed, it was not the worst greenhouse emitter in the world, because it has one of the highest percentages of natural gas consumption in the energy balance throughout the world. Let me emphasize that the countries of the former USSR demonstrate twice as much energy consumption per capita in comparison with the countries of the Eastern Europe. But at the same time, the greenhouse equivalent emission is not as high as we would expect if this structure of primary energy consumption was the same as Eastern Europe. Among the countries of Eastern Europe and the former USSR, Ukraine and Russia are approximately average producers of greenhouse gases. Poland and Czechoslovakia are much dirtier, and Romania, Bulgaria and Hungary are more or less cleaner producers of energy.

In conclusion let us consider the strategy of energy development in Eastern Europe and in the USSR. The decay of the USSR gave birth to a wide crowd of economic debris and the Eastern European countries are extremely concerned with their energy future. For a long period of time, their energy system had been developing as a satellite of the Soviet energy systems, and Soviet policy stimulated the development of energy intensive branches of the economy. After the export of subsidized energy resources from Russia stopped, those countries were in troubled waters, because the amounts of qualified energy resources like natural gas and oil and petroleum products dropped, and correspondingly over the next two years all of them suffered a drop in gross domestic product and national income of about ten to 15 percent.

There is an even worse situation in the former countries of the Soviet Union, the Ukraine, the Baltic countries, and Byelorussia, have had a the drop of national income during two years of about one third. Seventeen percent in Russia last year, 17.5, maybe 18 percent this year, or about 35 percent drop of national income in two years. The drop of energy supplies in the Eastern Europe countries, former republics of the former Soviet Union, brought those countries to the edge of economic catastrophe. For example, in Latvia, which has scarce domestic energy resources, industry is on the edge of stoppage, because there is no natural gas, there is no oil nor petroleum there.

So what is the strategy? I would say the strategy is survival. There is no working strategy in any way related to CO_2 emission or something like that. That's just high priority for the next century. Now all the strategy for the Eastern European countries and the former Soviet republics is to survive. Fortunately for the former Soviet republics which have a high share of natural gas supplies in the energy balance, this strategy aimed at survival, is not on a collision course with greenhouse gases emission. That is the only bright spot in the economic picture of contemporary energy development of the Eastern European countries and the former Soviet republics.

ENERGY USE IN INDIA: IMPLICATIONS OF CONSTRAINED GREENHOUSE GAS EMISSIONS

Ajay Mathur

Tata Energy Research Institute
New Delhi, India

India is one of the largest countries in the world in terms of both area and population, but one of the poorest in terms of per capita GDP and per capita energy consumption. Traditional energy forms (biomass, animal dung and agricultural wastes) constitute a large energy source in India, as in other developing countries and in the industrialized countries when at the same stage of development. As far as commercial fuels are concerned, coal is the pre-dominant energy source. Petroleum accounts for about one third of the total commercial energy consumption in the country and is growing rapidly. Natural gas has, as yet, a minor share, but will probably be a major fuel in the future. The share of biomass in the total energy mix has been decreasing steadily.

Over the past forty years, the total amount of biomass used for energy is estimated to have increased from about 90 Mtoe (million tons oil equivalent) to about 112 Mtoe. However, at the same time, commercial energy consumption has gone up from 19 Mtoe, as shown in Figure 1. Consequently, though the total energy intensity (i.e., the total energy consumed per unit of GDP produced) of the Indian economy has been declining (Figure 2), the commercial energy intensity of the economy (commercial energy consumed per unit of GDP produced) has been increasing.

As a consequence of this shift in energy-mix, carbon dioxide emissions have been on the rise: increasing from about 55 Mt of carbon to about 240 Mt carbon during the same period (Figure 1). This has had major repercussions on the overall economy of the country as it has implied major investments in the energy sector, as well as major foreign exchange outflows. Investment in the commercial energy sector (coal, petroleum and electricity) has grown from 1.5% of total national investment (and 0.175% of GDP) in 1950 to 12% of total national investment (and 3.25% of the GDP) in 1990. Oil imports accounted for over 40% of the total foreign exchange outflow in the late seventies/early eighties, but this dropped to about 15% by the late eighties as Bombay High production increasingly met local demands. However, with stagnating domestic oil production and increasing demands, oil imports are again on the rise and currently account for about 25% of foreign exchange outflow. It is expected that these foreign exchange requirements will severely constrain the economy.

Global Enery Strategies: Living with Restricted Greenhouse Gas Emissions, Edited by J.C. White, Plenum Press, New York, 1994

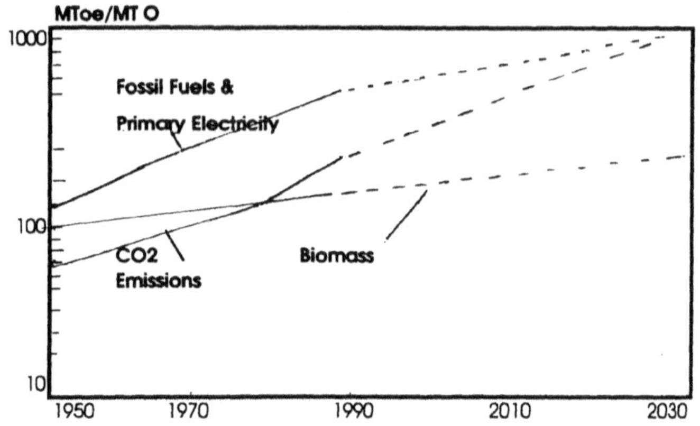

FIGURE 1. ENERGY USE AND CO_2 EMISSIONS - PATTERNS IN INDIA

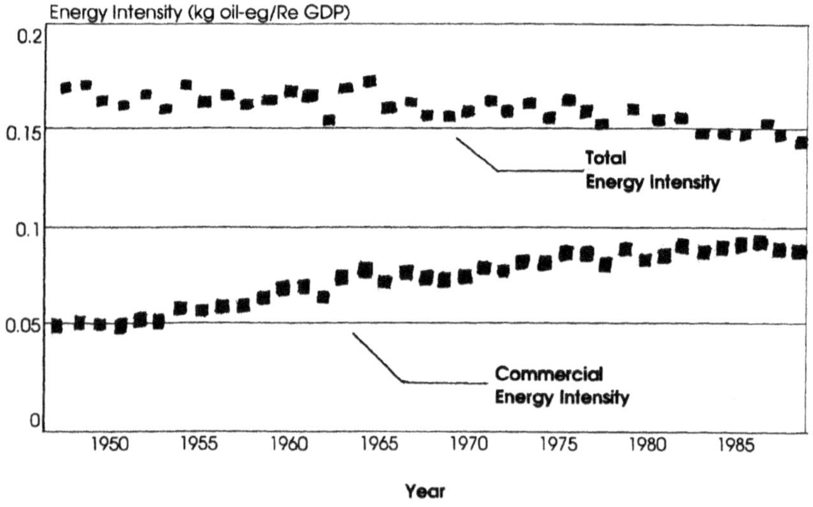

FIGURE 2. ENERGY INTENSITY OF THE INDIAN ECONOMY

SOCIO-ECONOMIC DETERMINANTS OF ENERGY CONSUMPTION

The domestic sector is the largest consumer of energy in India, and over 80% of the energy used in this sector is supplied by non-commercial sources, principally firewood, agricultural wastes and animal dung. Though non-commercial fuels dominate in both rural and urban households, there is a broad movement towards commercial fuels with increase in income, and more so with urbanization. Urban areas use 14kg of biofuel per capita per month, as compared to 29kg per capita in rural areas, with urban households substituting commercial fuels like coke and kerosene, for crop and animal wastes. Firewood, surprisingly, played an important role even in urban areas: it met more than half of the total gross energy needs of urban homes, similar to the pattern in rural households. The major distinction in the use pattern in rural and urban areas is seen across income levels. In rural areas, no significant change in use patterns is observed across income levels; in urban areas, however, the commercial/non-commercial mix changes from 1:3.7 in the lowest income category to 1:1.7 in the highest income category. Furthermore, useful (or delivered) per capita energy in rural areas is less than in urban areas, and the gross per capita energy consumption is higher in rural areas. A similar conclusion is also true across income levels. The low efficiency of about 8% of combustion of biomass fuels in traditional cookers is the prime reason for this anomaly. Inadequate fuel supply and lack of purchasing power in rural areas are the other major causes of the imbalance.

Daily per capita energy consumption in rural and urban households is approximately 350 and 430 kcals, respectively (ABE, 1985). These consumption norms are low by international standards, and are expected to rise to 520 and 650 kcals per capita per day by 2004-05. An important aspect of the consumption pattern is that over 65% of the biomass energy used in rural areas is collected. This places obvious constraints on fuel switching as a vast majority of the population does not have the financial resources to switch to commercial fuels even if they are available. However, the percentage of households dependent on collected biomass is expected to decrease with economic development since more households would have the economic ability to purchase commercial fuels, and since increased value of time would encourage a shift from biomass collection to economically more rewarding activities.

Since the average efficiency of traditional cookers burning biomass fuels is 8% (ABE, 1985), in order to promote energy conservation, the government of India launched a National Programme on Improved Chulhas (cooking stoves) in 1983 under which some 4 million improved stoves were installed during the period up to 1991. Preliminary results of an evaluation of the program found that about 65% of these improved stoves were in use, and resulted in fuel savings of about 25% over traditional stoves (Joshi, 1989). A similar program has been launched to promote biogas plants at the domestic and community level. A total of about 1.5 million plants have not been installed; while about 800,000 domestic plants are in use, community-sized plants have been very successful primarily because the sharing of both gas and costs has not been equitably arranged.

Electricity consumption in the rural household sector is insignificant at present. Electricity shortages have prevented the higher-income groups in rural areas from adopting it as a major energy source and consumption for the highest income is about 0.40kWh per capita per day. On the other hand, per capita electricity consumption in urban areas ranges from 0.3kWh per day for the lowest income group to 3kWh per day for the highest income group. Trends indicate that a doubling of income leads to a 75% increase in electricity consumption, with a saturation consumption level of about 5kWh per capita per day. The total commercial energy requirement in the household sector shows a very strong correlation with GDP. Electricity and petroleum consumption increase at rates which are greater than the rate of GDP growth (ABE, 1985).

The industrial sector consumes all commercial sources of energy: electricity, coal and oil. In general, the consumption of all fuels is increasing rapidly, though energy intensity is declining gradually as shown in Figure 3. For example, in the case of electricity, the ratio of consumption with respect to value-added in the industrial sector increased from 1.68 to 2.81 during the late 1960s on account of the substitution effect due to intensification of electricity use in the industrial sector. After that, the ratio declined, with an average value of 1.64 over the last 25 years; during the late eighties, it was around 1.48 (TERI, 1992a).

The major consumers of coal in the industrial sector are cement, fertilizers, paper, textile and brick-making. The intensity of coal use in industry has been declining gradually owing to substitution by oil and gas, as well as due to upgraduation of processes and coal utilization technologies. During the 1980s, constraints in supplies of coal to industry encouraged shifts to other fuels, primarily fuel oil and gas as indigenous supplies came onto the market (the elasticity of coal consumption was 0.64 with respect to value-added in industry; down from 0.8 in the 1960s) (TERI, 1992a).

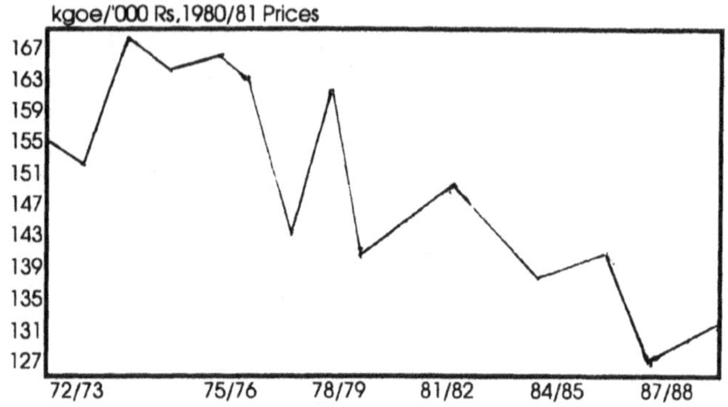

FIGURE 3. COMMERCIAL ENERGY INTENSITY IN THE INDUSTRIAL SECTOR

Oil and gas have been growing rapidly as industrial fuels, driven mainly by coal shortages and gas availability. Oil and gas utilization intensity decreased during the 1970s and early 1980s, and is steady at about 0.044kg per rupee of value-added.

The transport sector in India has been marked by rapid growth, with an increasing share of road transport in freight movement. The bulk of energy consumption in this sector is in the form of petroleum products. Coal use is expected to become insignificant by the turn of the century as the last coal locomotives are phased out. Because of these fuel drifts, the energy intensity of the transport sector has been decreasing (Figure 4), but energy consumption (of electricity and petroleum products) in the transport sector has exhibited a quadratic growth with respect to GDP (TERI, 1992a).

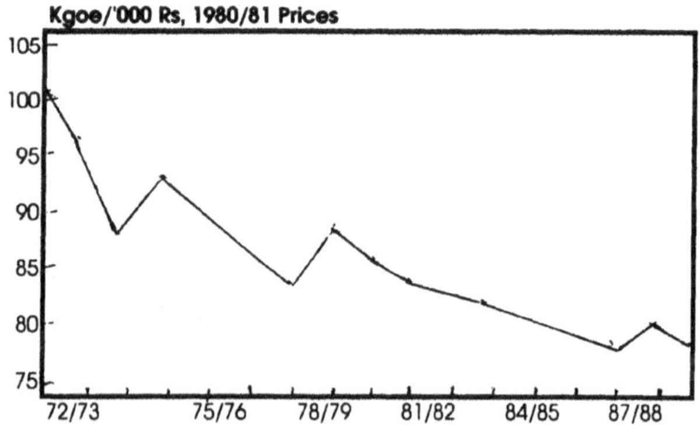

FIGURE 4. COMMERCIAL ENERGY INTENSITY - THE TRANSPORT SECTOR

PROJECTIONS FOR THE FUTURE

Projections over the next forty years indicate that primary energy consumption (including biomass) would increase by about 50% in the year 2000, and quadruple by 2030 (Mathur, 1991). Indigenous coal production is expected to remain as the dominant energy source in the future. However, petroleum demand would far outstrip indigenous petroleum supply. Domestic oil production is expected to meet only 50% of the oil demand in 2000, and less than one-third in 2030 (TERI, 1992b).

Figure 1 indicates the expected increases in energy supply and carbon emissions over the next forty years. In spite of the expected quadrupling of energy supply over that period, the annual per capita energy consumption and CO_2 emissions (of 0.62 Toe and 0.45 T carbon respectively) would still be far lower than present international averages (Figure 5).

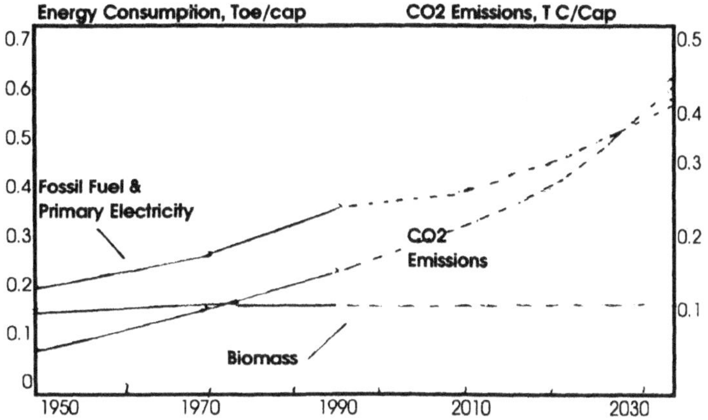

FIGURE 5. PER CAPITA ENERGY CONSUMPTION AND CO_2 EMISSIONS IN INDIA

LIMITATION OF CO_2 EMISSIONS: OPPORTUNITIES AND CONSTRAINTS

The low per capita figures, however, do not imply an inability to limit the increases in energy supply (and in carbon emissions). There are a wide range of technological options that are available for reducing CO_2 emissions. Figure 6 illustrates a capital cost curve for CO_2 emissions abatement in India: the total abatement potential (through mitigation and enhanced sequestration) is more than the current level of CO_2 emissions (Mathur, et al., 1992). Many of these options, including CO_2 limitation options, that are desirable in India for many reasons other than global-warming abatement. Three such major strategies are: afforestation, energy efficiency enhancement, and development and deployment of renewable energy technologies. Each has been individually identified as a priority area, yet capital constraints have limited their impact. For example, there is a stated objective of afforesting 30% of the country's land area. This would be about 109 million hectares (Mha), and currently the area under forest cover is about 64 Mha. In the mid-1980s, afforestation was high on the government's agenda, and an afforestation target of 5 Mha per year was adopted. Even this relatively modest target could not be achieved, and only 1.75 to 2 Mha could be afforested annually in the late 1980s.

In this perspective of the net shortage of capital in the economy, additional program for CO_2 limitation would inevitable flounder. These programs would receive some financial support, but only at present levels of funding. Major limitation programs would, therefore, necessarily require support from external resources. Such support, on an incremental cost basis, could be linked to a particular CO_2 limitation program within the perspective of an overall target at the national sectoral level.

It is the author's estimation that the rate of growth of CO_2 emissions can be reduced by 20% if appropriate macroeconomic restructuring is carried out and adequate capital and technological resources are made available. The central problem of Indian development is that a vast proportion of the population - certainly in excess of 50% - is either unemployed or employed in marginally-productive jobs. Consequently, per capita net resource creation is low, and in spite of the high savings rate in the country (in excess of 24% of the GDP), the per capita annual domestic investment is currently a meager Rs.2,700 (GoI, 1992). This is invested in generating employment, as well as for providing amenities (food, shelter, energy) to the population that is unemployed or underemployed. The emphasis of government policy has been on rapid employment generation - and creating a job in the organized sector now costs more than Rs.50,000. It is therefore, politically unacceptable to

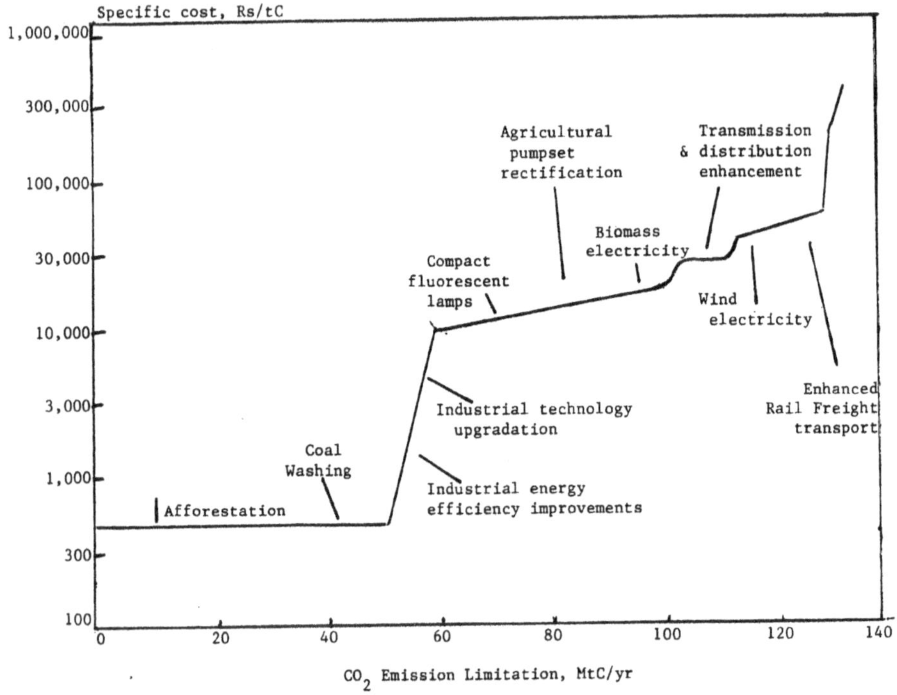

FIGURE 6. COST CURVE FOR CO_2 LIMITATION STRATEGIES IN INDIA

NOTE: Only major strategies have been identified in this figure.
A total of 29 strategies are included in the analysis.

move capital from programs for employment generation to those for limiting CO_2 emissions. This capital shortage has long stymied efficiency measures, eg. efficient operation of electric utilities, and energy conservation in industry, even though most energy conservation measures are negative-cost strategies over a period of time.

The estimated 20% reduction in the rate of growth of CO_2 emissions is expected to cost about Rs.75 billion, or about 0.53% of the annual gross domestic investment. Interestingly, the shift in domestic investment for reducing the rate of CO_2 emissions growth varies inversely as the per capita GDP of the country: Table 1 illustrates that the, "pain of adjustment," to Bangladesh is a lot more than to Brazil, though the limitation target for Brazil is five times that for Bangladesh. It is therefore, politically difficult to restructure investment priorities as to favor abatement in poor countries such as India.

TABLE 1. IMPACT OF CARBON DIOXIDE EMISSIONS LIMITATION ON THE ECONOMICS OF DEVELOPING COUNTRIES

	1990 Emissions (MT C)	Expected Emissions in 2000 (MT C)	Target Reduction in Annual Emissions by 2000 (MT C)	Annual Cost of Emissions Limitation (Million $)	Annual Gross Domestic Investment (Billion $)	Shift In Investment (%)	GDP Per Capita (%)
Bangladesh	4.24	8.87	1.18	17.7	2.42	0.73	180
India	160	248	20	300	56.24	0.53	340
China	588	964	89	756	150.12	0.5	350
Thailand	37.8	74.4	0.2	65	21.6	0.3	1220
Brazil	66	94	6.3	95	70.21	0.135	2540

Data Sources: AEI (1991); World Bank (1991)

Limitation Target: Reduction of 20% In Annual Rate of Growth of Carbon Dioxide Emissions during the 1990's

Finally, it is imperative that individuals in organizations involved with energy production or consumption develop the capability to enhance and sustain energy efficiency (Bell, 1990). For this, a broad range of macroeconomic policy initiatives (e.g., rational pricing of energy, and evolving set of energy standards, etc.) are required, as well as a deepening of the scope of technology transfer to enable the recipient firms to meet the changing economic and technical requirements necessitated by the threat of global climate change.

REFERENCES

ABE, 1985; Towards Perspective on Energy Demand and Supply in India in 2004/05, Advisory Board on Energy, Government of India.

AEI, 1991; Global Warming: Mitigation Strategies and Perspectives from Asia and Brazil, eds. Pachauri, R.K. and Behl, A., Asian Energy Institute and Tata McGraw-Hill, New Delhi.

Bell, M., 1990; Continuing Industrialization, Climate Change and International Technology Transfer, SPRU, University of Sussex, Brighton.

GoI, 1992; Economic Survey 1991-92, Ministry of Finance, Government of India.

Joshi, V., 1989; Rural Energy Demand and the Role of Improved Chulhas, In Energy policy Issues, Vol. 4, TERI, New Delhi.

Hossain, J., and Raghvan, K., 1001; Potential for Wind Farms in India, In Innovation in Indian Power Sector, TERI, New Delhi.

Mathur, A., 1991; The Greenhouse Effect in India: Vast Opportunities and Constraints, In Energy Policies and the Greenhouse Effect, Dartmonth, Aldershot.

Mathur, A., Gupta, S., Khanna, N., 1992; India, In Global Warming: Collaborative Study on Strategies to Limit CO_2 Emissions in Asia and Brazil, eds. Pachauri, R.K. and Bhandari, P., p. 65-89, Asian Energy Institute and Tata McGraw-Hill, New Delhi.

TERI, 1992a; TERI Energy Data Directory and Yearbook, TERI, New Delhi.

TERI, 1992b; "Fuelish" Trends and Wise Choices: Options for the Future, TERI, New Delhi.
World Bank, 1991; World Development Report 1991, Oxford University Press, New York.

SOUTH AMERICA

Russell Sturm

International Institute for Energy Conservation

You have heard a lot of the concepts that I'm presenting in this paper. I want to talk about the direction that the South American and Latin American regions are going to move in the energy front, which will be an essential part of their development.

IIEC, the International Institute for Energy Conservation, is a non-profit organization, based in Washington, D.C. We have offices in Bangkok, Thailand and Santiago, Chile, where we take a model approach focusing on these countries. We make commitments in these countries, to help develop the regions in which they're situated, using energy efficiency as a primary development tool. We expect to open an office in Brussels later next year. The next regions of our focus will be Africa and Central and Eastern Europe.

We work to facilitate the design and implementation of least-cost utility planning, promoting an integrated resource planning approach to energy development. I work with the private sector to make certain there is a capacity in these markets to deliver energy efficient products and services. In addition, we have an information project that supports the various activities of the organization. We serve as the secretariat for the Global Energy Efficiency Initiative. We have a formidable and growing transportation program, and we work with the multilateral development banks to integrate energy efficiency as a development focus.

I am going to respond in part to some of the concepts that Mr. Mathur has laid out. He writes about capital limitations providing an impediment to energy efficiency on a global scale. Well, there's several reasons why we focus on energy efficiency as a means towards development. One has to do with the importance of greenhouse gasses, which is the focus of this conference. Currently, in the developing countries, there are 5.2 billion metric tons of CO2 emissions because of energy consumption. The projections on a base case scenario show that by the year 2025, that will more than double. With a conservative efficiency component to energy development, we can get 25 percent savings.

I want to respond to Mr. Mathur's reference to the capital side of this. The other component of the imperative to utilize energy efficiency to the maximum is the capital shortage. Energy efficiency is a capital efficient investment. The World Bank projects that in the remainder of the '90s, on a business-as-usual scenario, there will be a requirement for one trillion dollars in capital to meet the electricity sector development needs of the developing world alone. The wealth of capital available for that part of the world for energy development totals about $230 billion, which means more than a 75 percent shortfall in a business-as-usual scenario. This means either you have the capital shortages convert to stoppages in economic development, or you have to use your capital more efficiently. Energy efficiency is one means to achieve that end, in fact, a primary means.

The U.S. and other OECD countries currently represent about half of the primary energy consumption in the world, the countries I'm talking about define the other half. What's more important is, when you look at the growth rates in the developing countries compared to the U.S. and other OECD countries, the imperative becomes clear and supports the essential importance of focusing on the developing world.

This translates into the electricity sector. The concepts that I'm laying forth here today apply not just in the electricity sector, which is an easy one for me to talk about because we can focus on it, and it is in transformation.

When we look at it on a regional basis within the developing countries, it mirrors the approach that IIEC has taken. Our first office was opened in Bangkok, Asia, and represents the bulk of the growing demand projected. Our Latin American office was opened in Chile last year, next is Eastern Europe, and following that is Africa. Ms. Bleviss made reference to Africa in her preceding paper. Given the energy usage from traditional means that Mr. Mathur makes reference to, the significance on a global scale of Africa is not great, especially relative to CO2 output. The growth figures, however, point to its importance in the years to come, and the importance of energy efficiency as a way to get Africa on the development track to improve quality of life.

Now, in Latin America, which is the focus of my talk, energy intensity is increasing relative to gross domestic product. All of this paints a picture which leads one to think, where do we go? We're talking about taking a less painful solution approach. How do we face the future when we're swimming upstream? Is there an attractive means to get up that hill? The answer in part has to do with transferring policy ideas.

Let's look at North America for one second, if I can step back. Why is it that utilities like Southern California Edison pay builders $30 per square foot to install super windows, which the builders would not otherwise install? Why is British Columbia Hydro providing incentives, ten dollars per horsepower, for motor replacements? Along the same ideas, last year, U.S. electric utilities paid customers half a billion dollars to install efficient lighting and two billion dollars for all their demand side management investments. One of the prime examples of this is Pacific Gas and Electric, which is in its long-term planning scenario going to get 75 percent of their added energy services from demand side management. They're essentially out of the business of building power plants. Finally, in North America, why did the consortium of some 25 utilities put together a pool of $30 million to create an inducement for refrigerator manufacturers to push the technology ahead, in terms of efficiency? Why is it that IIEC has had success in Thailand, to the point where the government there is embracing demand side management as a fundamental component of an integrated resource plan, where they are using efficiency as a means to expand energy services?

The answer is simple. Demand side management, investment in energy efficiency, is profitable if the institutional mechanisms are in place to allow the market forces to dictate investment. Compare a typical investment that commercial builders are making in energy efficiency, getting a return on investment of 33 percent, against alternative investments for that company. Treasury bills yield eight percent, the Dow Jones average is 16 percent.

What sort of paradigm allows these sort of investments to come about? It's one where you include the potential resources as you expand energy resources to include demand side investments. When you do that, potential investments in commercial lighting result in expanding energy services from the electrical sector at the cost of .7 cents per kilowatt hour. You can then include each additional demand side resource that's available, and determine at what level you need to continue to expand your resources. This sort of analysis is what led Pacific Gas and Electric to devote 75 percent of their future resource acquisition to demand side management. Compared to an out-of-state coal plant charging ten cents a kilowatt hour, it doesn't compete.

There is a similar supply study done for Sweden, they list various resources that the utility in Sweden has integrated into its resource plan, including new home construction, conversion of large heat pumps and even the cost of an existing hydro plant.

What you have traditionally is consumers with very high personal discount rates, leading them to make uneconomical decisions. Mr. Mathur had some figures which I think are low compared to a lot of data I've seen, which translate into consumers not making purchases that have paybacks under half a year. At the same time, utilities traditionally do not have incentives to look beyond a stack of resources just involving power plants. So the utilities' investment perspective is about a 30 year horizon; it will take 20 or 30 years to amortize this power plant investment. The consumer would have to face a price of 52 and a half cents per kilowatt hour before he would make an investment in a more expensive, efficient refrigerator. By comparison, the utility or whoever would make this investment is actually acquiring additional power at two and a half cents a kilowatt hour.

The result of regulatory reforms in North America is that this sort of approach has been taken, where the utilities look for mechanisms to invest in the end use. With this sort of paradigm shift, the global market for energy resources would stack energy efficiency as a competing resource in each of these regions.

Now, how do you go about doing this? What we're talking about is looking at the capital costs of building one compact fluorescent factory that has an output energy savings the equivalent of a 700 megawatt power plant. Seven and a half million dollar investment in the compact fluorescent plant translating into a one billion dollar power plant. That's the capital savings I'm referring to. What we're talking about is comparative investments in the coal plant versus the low emissivity windows-coating factory.

In California between the years '73 and '86, energy usage increased by nine percent, population grew by 31 percent and the gross state product grew at 58 percent. What are the institutional impediments to this? This is a philosophy that has to change. The Corps of Engineers mentality looks at the need for expanding energy resources as a one-shot deal, a one-resource option. You build the dam and build the power plant, and you don't consider the other alternatives. Institutionally, they may not be set up to do that yet. U.S. Department of Energy spending for research on conservation is very low relative to traditional fuels. The other International Energy Agency member nations have a very similar breakdown.

The World Bank, which is the key lending facility for the developing world during the decade of the 1980s, had total energy lending of $37.75 billion. Less than one percent went for efficiency, and of that, 58 percent was dispersed for real projects. What we often hear from the development agencies is that it's difficult to find an efficiency project to invest it. It doesn't fit the structure of how they do business. They don't quite understand how it goes. This is a problem we're working on. How do you create a bankable project? How do you create the institutions in developing countries to make efficiency a viable and usable resource?

Well, one such structure is to involve the utility in selling energy services. I don't want to focus on one technology, because we're talking about a whole range of them, motors, building products, lighting products, controls. One example is a compact fluorescent lease arrangement. A utility can make a large purchase of light bulbs, for example, compact fluorescents. It can get a preferred price. They can provide market information. They can also provide financing, so that users pay back the cost of the light bulb on their utility bill at a rate which is less than the amount of savings that they actually generate, so there's a net gain to the consumer on a monthly basis. There's a lot of advantages to that.

Those advantages are now going to be felt in Mexico. Mexico is a country of 90 million people. Per capita income in 1989 was about $1900; 24,000 megawatts is the electrical capacity of the country. It is growing at seven percent a year and is expected to reach 75,000 megawatts by the year 2010. The same capital constraints I made reference to earlier face

Mexico. In addition, they burn a lot of oil for their electricity. There's a strong interest in the country in reducing usage of oil so they can export it.

The IIEC suggest that the Mexican energy resources, be increased by 800 terrawatt hours per year and use efficiency potential. The benefits from Mexico adopting demand side management as a component in integrating all the resources available to them for electricity development are many-fold. Energy efficiency is very labor intensive in terms of dollars input, the amount of jobs created. It frees up oil, it has environmental benefits.

I'm going to discuss in particular one project. It's called Ilumex. It's the first large-scale demand side management program being undertaken with World Bank funding. It's being undertaken in two cities in Mexico, Monterey and Guadelajara. It's a $20 million dollar project. It involves the purchase of one and a half to two million compact fluorescent lamps. It's about ten percent administrative costs, and ten percent capital costs for purchasing the lamps. We put a budget together for monitoring and evaluation so we can learn from this project. The intent of the Mexicans is to expand it. The benefits of an individual compact fluorescent are manyfold from an environmental perspective.

As I mentioned, we are starting in two cities. It's a revolving fund so it will be expanded for the nation as a whole. There's a commitment from the utility to do that. The emissions being saved from this $20 million investment which regenerates, is a zero investment after three years. It is in the range of saving 117 to 156 kilotons of carbon, sulfur in the 13 to 18 kiloton range, and NOX savings at about a kiloton, all of which is achieved at a negative cost.

This comes about because the fuel saved is 265 barrels a year, deferring a peak capacity of 123 megawatts. As I mentioned, the program will be expanded Mexico-wide, with fuel savings of 4.06 million barrels a year.

Motors are an example of another technology. Two thirds of electricity in developed countries -- and it's potentially higher in developing countries -- is consumed by electric motors in electric processes and building operations. The motor's annual operating costs are ten times larger than the first cost of the motor. We're talking about paybacks for replacing motors in less than one year.

Howard Geller at the American Council for an Energy Efficient Economy has done extensive work on the efficiency potential in Brazil. The scenario is that, with the relatively conservative investment in efficiency, you can actually cut in half the amount of electricity consumption growth by the year 2010.

Finally, I will discuss a similar scenario for Chile, where we have just begun to work. This is based on a very preliminary end-use assessment, which is the first step in doing a demand side management initiative. You have to understand what your resources are, just as you have to understand what sort of oil is available to you as a country. You need to understand how you're using energy currently, and what sort of opportunities can be mined in this end-use manner.

In Chile, we're talking about power savings, with a high conservation scenario on the order of 3,000 megawatts. That compares against projected forecast for new power consumption, again, very consistent with what Howard Geller found in Brazil, of about half. That is to say, you can cut your gross in half and expand resources at the rate required for development.

REDUCING ENERGY-RELATED GREENHOUSE GAS EMISSIONS: A CANADIAN PERSPECTIVE

Bunli Yang

Ontario Ministry of Energy

ABSTRACT

Canada is a large, northern country with a sparse population and an industrialized economy. Its economy has been highly energy-intensive, taking advantage of a substantial base of natural resources. Canada intends by 2000 to keep its national greenhouse gas emissions below 1990 levels while embarking on economic recovery. This will be a difficult challenge for Canada's energy economy, on both the supply and demand sides. The methods are the same as elsewhere -- greater efficiency, substitution of zero- or low-carbon fuels, new technologies, and better pricing. Setting priorities and making choices will involve a complex coordination of efforts by federal and provincial governments in partnership with business and individuals. Competitiveness of Canadian industry and assistance to developing countries will be major factors. Innovative policies are needed for industrialized countries to improve the efficiency of energy supply and use at home, and abroad in the developing countries.

INTRODUCTION

Canada has a population of about 27 million people on a land area larger than the USA or China. Canada has a rich base of natural resources, and its economic structure in large measure has depended on the production of commodities, from aluminum and nickel to barley and newsprint. Canada's production of many commodities -- and its use of energy -- on a per capita basis are among the world's highest.

Canada's energy mix is diverse, about two-thirds being fossil fuels, with transportation accounting for most of the use of oil. Electricity supply is dominated by hydropower, about two-thirds, with the rest split nearly equally between nuclear and coal. There is currently very little oil- or natural gas-fired electric power. In total, with about 0.5 per cent of the world's population, Canada is responsible for about 2.5 per cent of global CO_2 emissions.

Canada's large size and cold climate lead to heavy dependence on transportation services and on heating. And the production of resource commodities is heavily energy intensive, so that Canada's use of energy per unit of gross domestic product (GDP) is high when compared with other industrialized countries.

In terms of dealing with climate change, the rate of change of energy intensity of the economy, not the level itself, will be the main indicator of success. Canada's reduction of its energy/GDP since 1970 has been comparable with other OECD countries. Even tracking the change of energy/GDP over time gives only a rough indication of improvements in energy efficiency because structural changes in the economy also alter energy/GDP. Breaking out the changes over time in energy/GDP to account for population changes, fossil fuel use, and primary energy mix can give a deeper insight to whether an industrial economy is making structural adjustments to reduce greenhouse gas emissions. For example, steady population growth during an economic recession might lead to an apparent improvement in energy/GDP, but would not be the result of a policy to improve energy efficiency. Similarly, simply tracking CO_2 emissions also gives little indication of whether structural changes in the energy economy are leading to reductions in emissions. An example of how to understand the structural reasons for CO_2 emissions:

$$CO_2 = \frac{CO_2}{\text{fossil fuel}} \times \frac{\text{Fossil Fuel}}{\text{Primary energy}} \times \frac{\text{primary energy}}{\text{secondary energy}} \times \frac{\text{secondary energy}}{\text{(GDP) Gross Domestic Product}} \times \frac{\text{GDP}}{\text{population}} \times \text{population}$$

CANADA'S INTERNATIONAL COMMITMENTS

As a signer of the Framework Convention on Climate Change, Canada has committed itself to:

- emissions reductions, with reporting on policies and measures;
- assistance to developing countries through funding and technology transfer;
- reporting on inventories of emissions, national programs, and assessments of progress;
- continued technical and socio-economic research; and
- expanded public education and training.

In addition to these commitments under the Convention, Canada has announced its "Quick Start Agenda", which includes:

- hosting an international meeting on developing the methodologies to calculate emissions by sources and removals by sinks;
- sponsoring a meeting with the Global Environment Facility of the World Bank to mobilize funding for country studies;
- developing with other countries a workplan for the International Panel of Climate Change (IPCC);
- issuing a national report on Canada's plans and policies by June 1993; and
- helping to build capacity in developing countries for country studies, planning, and reporting requirements.

CANADA'S AGENDA AT HOME

Canada's Green Plan, issued in late 1990, contains a range of initiatives and commitments for environmental improvement. A key target is to stabilize national emissions of CO_2 and other greenhouse gases, not controlled by the Montreal Protocol, at 1990 levels by the year 2000. This national goal, during an economic recovery from the painful recession of the past two years, is a very substantial challenge.

The national target will require an unprecedented level of coordination and cooperation between our federal and provincial governments. In Canada's system of governance, the provincial governments have ownership of natural resources and have extensive powers over regulatory, tax, and economic matters. Thus, many of the implementation responsibilities to reduce greenhouse gas emissions fall to the provinces.

Furthermore, the mix of energy supply and use varies widely across the country. For example, nearly all of Quebec's electricity is hydraulic; half of Ontario's electricity is nuclear; and virtually all of Alberta's is coal-fired. What is cost-effective in one jurisdiction may not be relevant in another.

The division of powers, together with the wide diversity of regional energy economies across the country, means that developing a domestic strategy and plan of actions will require wide, intense consultation and cooperation among governments.

In several provinces, the recognition that many environmental issues related to air quality are driven by energy use has led to joint efforts by energy and environment departments to develop "clean air strategies" that emphasize prevention of emissions at source rather than control at end-of-pipe. Thus, the cost-effective ways to deal with the issues of acid rain, urban smog, and greenhouse gases tend to be improvement of energy efficiencies and substitution by renewable or low-emitting sources of energy.

Technology specifications and standards (or what is often called command-and-control) will remain important and will be expanded. But bringing about major changes in Canada's patterns of energy use by businesses and citizens will likely also depend heavily on marketplace incentives and signals.

Some economic instruments are already in place, originally intended to serve other purposes, but which stimulate energy conservation all the same. One example in Ontario is the combined federal and provincial excise tax on gasoline -- the tax is currently about \$0.24Can/liter, or about \$0.75US/gallon. Another example is Ontario's Tax for Fuel Conservation, a "gas guzzler" tax on new cars, based on fuel economy, which gives rebates to buyers of the most efficient new cars.

Going beyond the current regulatory, tax, and program efforts of governments is very tough sledding during one of the worst recessions in memory. Competitiveness of Canadian industry is a key concern. Inevitably, the business community is wary of new pricing signals -- regarded as taxes -- whether on energy or on new capital investments to capture "lost opportunities" for improved energy efficiency.

The main thrust now is to identify and adopt measures and actions that are directly cost-effective at current market prices. Whether and how to incorporate externalities into decisions on what kinds of energy supply investments to make, or even into energy pricing, are beginning to be addressed but only for such issues as acid rain and smog, not yet for greenhouse gases. Altering market prices for energy can severely affect industrial competitiveness, and damage functions for greenhouse gas emissions will be very difficult to estimate because of the global, non-local nature of their impacts, first on radiative forcing in the atmosphere and then on any resulting changes in climate and sea levels.

POSSIBILITIES FOR GLOBAL COOPERATION

Canada's "Quick Start Agenda" is a first step in turning the Framework Convention into concrete action, both at home and abroad. Tracking real progress in the industrialized countries such as Canada will require sophisticated analysis of changes over time of emissions and their causes, as discussed earlier. The industrialized countries may be a large part of the problem now, but they are not alone.

In the developing world, rates of population growth and of economic growth needed to raise living standards pose the challenge of rising greenhouse gas emissions and, therefore, of directing investment capital flows into improved energy efficiency in the developing countries. The Global Environment Facility at the World Bank is a pilot attempt to provide such capital assistance, but it remains a small-scale, yet centrally-directed approach to project assistance.

Allocating responsibility to provide monies for technology transfer and setting up mechanisms to direct the spending are difficult enough when considering government-to-government efforts. The larger task must be to influence the much more substantial flows of private sector investment capital. Such a conclusion implies that systems of incentives are needed.

A different approach to capital assistance for improving energy efficiency in developing countries would be to establish incentives within a global policy market -- a market in which governments are the actors and also set the rules and incentives for private sector behavior -- for providing capital and seeking cost-effective opportunities to improve energy efficiency. The key recognition is that energy inefficiency anywhere contributes equally to the greenhouse gas problem; and the lowest capital cost investment opportunities to improve energy efficiency will be found in developing countries.

An example of such a system of "incentives to do well and to do good" is a system of *tradeable warming credits (TWCs)*, beginning with the industrialized countries, and later including the developing countries. The OECD countries comprise less than one-sixth of the world's population but about half of the world's CO_2 emissions, so the TWCs strategy could start with the OECD and make major gains. The TWCs strategy for OECD governments:

1. Accept the responsibility of OECD countries to set up technology transfer funds to improve energy efficiency in developing countries.
2. Allocate annual funding responsibilities among OECD members according to the carbon content of energy used by each member country in a baseline year. For example, a charge of $0.50 US per tonne of carbon would raise about $1.3 billion US annually, with the US accounting for about half of the total.
3. Call these annual funding responsibilities "warming credits" and make them tradeable between OECD governments only. Energy users would not be involved. All trades would result in exchanges of revenues between OECD member governments.
4. Set a higher, excess charge of say, $5 US per tonne of carbon for any OECD country exceeding the TWCs it acquires either from its annual allocation or from any trading purchases from other OECD countries. These excess charges become that country's additional contributions to the technology transfer fund.
5. Set up a means to spend the revenues to improve energy efficiency in developing countries together with a mechanism to re-direct existing cashflows within those countries; for example, swaps of foreign-held debt for energy efficiency.

TWCs apply to governments, not to energy users, so that TWCs introduce no new pricing signals to energy users. Instead, the TWCs system would establish a framework of incentives for governments to undertake policies to improve energy efficiency, with the discipline of public scrutiny to track progress on the most direct measure of success: saving money. The excess charge, at least 10 times the TWCs charge, can be viewed as an avoidable carbon tax on OECD governments, avoidable through successful policies to improve their economies' energy efficiency.

STRATEGIES FOR IMPROVING ENERGY EFFICIENCY AND REDUCING CO2 EMISSIONS IN THE EUROPEAN COMMUNITY AND THE NETHERLANDS

Hans van Zijst

Royal Netherlands Embassy

INTRODUCTION

This contribution to the Proceedings on Global Energy Strategies is about three topics: the CO_2 strategy of the European Community, the CO_2 strategy of the Netherlands and the global strategies for the longer term. Documentation used for the preparation of this paper includes "The economics of limiting CO_2 emissions", issued by the European Community in 1992 (1) and "Climate Change", the Dutch whitepaper on CO_2 policies (2).

THE CO2 STRATEGY OF THE EUROPEAN COMMUNITY

Climate Change is a phenomenon which is caused by activities from every quarter of the globe. Its impacts will be felt on a planetary scale. The policy response to this phenomenon must therefore also be a worldwide one. No one country or group of countries acting alone can reverse the trend of global warming. However both ethical and economic arguments compel the conclusion that industrialized countries should take the lead in tackling this issue.

Ethical arguments because it is the industrialized countries who have contributed the most to causing the problem thus far. The EC's share in world CO_2 emissions is 13%, that of the USA is 25% and Japan 5%. The countries of central and eastern Europe and the Commonwealth of Independent States contribute about 25%. Less than 30% of the world's population is thus responsible for nearly 70% of anthropogenic carbon dioxide emissions.

And economic arguments because there will be clear first-mover advantages for those countries who are first to develop and apply sustainable energy systems.

The European Community and its Member States therefore have expressed their intention to play a leading role in international efforts to limit the impacts of human activities on the planet's climate system. To this end, the European Community has made a political commitment to the objective of stabilizing its CO_2 emissions at their 1990 level by the year 2000 in the Community as a whole. Adoption of the objective in the autumn

Global Enery Strategies: Living with Restricted Greenhouse Gas Emissions, Edited by J.C. White, Plenum Press, New York, 1994

of 1990 was intended to further global action at an early stage, a year and a half before the UNCED meeting in Rio.

In order to translate this political commitment into hard action, the Commission of the European Communities has proposed a strategy to limit CO2 emissions and improve energy efficiency in the Community. The general philosophy underlying the proposed Community strategy is one of "designing a cost-effective response to the risks of global warming, which at the same time can be perceived as being equitable at both the individual and the Member State level". The Commission also devoted attention to the question of the level of government at which measures should be taken - Community or Member State. Three fundamental principles were thus of critical importance in the development of the strategy: economic efficiency, equity and subsidiarity.

The strategy proposed by the Commission was discussed and endorsed in a joint Energy/Environment Council on December 13th, 1991. The Council recognized that, in order to reach the CO2 stabilization target in a cost-effective way, higher energy pricing through the use of fiscal instruments was likely to be needed to complement national and Community energy efficiency programs. The Council asked the Commission to devote further study to a number of questions and to make concrete proposals for implementing the strategy.

A mix of different policy instruments is required if the stabilization objective is to be attained in an economically acceptable way. Thus the strategy proposed by the Commission consists of three components:

1. the so-called 'no-regret' policies, i.e. policy measures that generate net benefits even without reference to the issue of global climate change;
2. a Community-wide carbon/energy tax;
3. in line with the subsidiarity principle, complementary national programs adapted to each Member State's specific circumstances.

The 'no-regret' component of the strategy consists mainly of measures designed to encourage the rational use of energy. They are of fundamental importance for attaining the emission stabilization target at a low cost or even at an economic benefit. The beneficial effects of the 'no-regret' measures will include such things as improving energy security, reducing emissions of pollutants other than CO2, encouraging more effective use of the transportation infrastructure and developing environmentally friendly means of transport. They will also enhance the competitiveness of European industry because they lead to the manufacture of products and the development of production techniques which are not harmful to the environment. Demand for such environmentally friendly products and techniques is growing by leaps and bounds worldwide.

Although the measures envisaged are expected to generate net economic benefits over the long term, there may be adjustment costs in the short term at both the macro-economic level and at the level of individual companies. This is particularly true in the case of the carbon/energy tax. Adherence to the principle of cost-effectiveness implies the need for the application of broad-based and Community-wide policy instruments using the market mechanism for reducing carbon dioxide emissions. In view of the transnational nature of the problem, and taking into account the important potential implications for competitiveness, the application of the subsidiarity principle leads to the conclusion that the structure of such policy instruments should be decided at the Community level. In this context, the choice of a Community-wide tax has the clear advantage that emissions are reduced where the costs of emission reduction are lowest. Various steps can be taken to ensure that the adjustment costs remain as low as possible. First of all, if the tax is introduced in a way that is fiscally neutral, the reductions in other taxes can contribute to mitigating any negative economic impacts. And second, adjustment costs will be reduced if the tax is introduced in a gradual and foreseeable way.

An alternative to the cost-effectiveness approach would be one based on the idea of reducing emissions where this would appear 'fair', rather than where the costs are lowest. However, such an approach would be very expensive. Instead, the aim for an equitable sharing of the burden can be dealt with by making use of the policy instruments appropriate for dealing with this equity aspect, both at the national and the Community level. It is very difficult to estimate the real costs of the adjustment process in each of the countries. However, should such costs prove to be disproportionately high in relation to the economic situation in a Member State, it would be possible to arrange financial support through various funds or Community financial instruments.

The Commission has made five concrete proposals for implementing the strategy, all of which are currently under discussion. These concrete proposals are intended to improve energy efficiency in the short term and to stimulate fuel substitution in the longer term. Energy efficiency improvements can generally be achieved within a shorter time horizon than fuel switching, which may necessitate changing the structure of energy supply. While it may be necessary to turn over end-use equipment in order to improve energy efficiency, and while this takes some time, changes in the supply structure usually involve even longer lead times. Considering the short time horizon until the year 2000, measures to improve energy efficiency appear to be more promising in achieving the stabilization target for 2000. In the longer term beyond 2000 fuel substitution might have a greater role to play in limiting CO2 emissions.

The EC has estimated that achieving the stabilization target will require that emissions of carbon dioxide within the Community (including the former East-Germany) will have to be reduced by 12% in 2000. This reduction is in addition to the normal gains in efficiency expected over the period and amounts to 335 million tonnes of CO2. The EC aims to achieve this 12% reduction through a series of measures, that are currently in debate:

a. a Community-wide carbon/energy tax, expected to contribute about 25% of the reduction needed, or 3 percentage points of the total 12;
b. the EC's SAVE program (Specific Actions for Vigorous Energy saving), which could contribute another 3 percentage points of the reduction;
c. a 1993 call for tenders focusing on the reduction of CO2 emissions in the framework of the THERMIE program (promotion of European energy technologies), accounting for 1.5 percentage points of the reduction;
d. activities in the context of the EC ALTENER program (to promote penetration of renewable energy sources), which will reduce emissions by one additional percentage point, and
e. complementary national programs which will have to make up the difference of about 3.5 percentage points.

THE CO2 STRATEGY OF THE NETHERLANDS

Since a very important element of the EC strategy for achieving its stabilization goal relies on the implementation of national programs, I want to say a few words about what the Netherlands is doing in this area. The Netherlands is one of the four EC countries (along with Germany, Denmark and Belgium) which has committed itself to a CO2 reduction objective that goes further than the Community as a whole. Although the Netherlands' contribution to worldwide CO2 emisions amounts to less than 1%, we still feel that the equity considerations that I mentioned earlier compel us to take action to control our emissions now.

Our objective for CO2 reduction was set out in the National Environmental Policy Plan issued in 1990. It requires us to reduce our CO2 emissions by 3 to 5 % in 2000 relative to the average level in 1989-1990. Given the current projections of how emissions may be expected to develop during the coming years, this means that emissions will have to be reduced by 41 to 47 million tonnes in 2000 compared to what they would have been in the absence of policy measures. About 42% of this reduction will have to be realized through energy conservation, 32% through fuel switching, 16% through measures in the area of traffic and 10% through waste prevention. As you can see, the biggest chunk of the total reduction will have to be achieved through more efficient use of energy. In 1990 the Netherlands set itself the task of increasing its annual energy efficiency improvement rate to over 2%, about twice the rate of improvement expected given market developments and government policies in place at that time. What does this 2% mean? It means that during each year in the period 1990-2000, the Netherlands will have to use 2% less energy to perform the same activities as in the preceding year. No small challenge!

The goal itself as well as the strategy for achieving it is derived from the 'no-regrets' principle. We share the EC's confidence that there are major first-mover advantages to be derived from improving the efficiency with which the Dutch economy uses energy. There will also be benefits flowing from the reduction in emissions of acidifying substances; these benefits will largely accrue directly to the Netherlands. But our neighbouring countries will share in them to some extent as well.

We have recently started discussions about the strategy to be pursued in the years after 2000. These discussions are still in a very preliminary stage, so it is not possible for me to present any conclusions. But you may find it interesting to hear about one of the ideas which is starting to circulate. This is the idea of an 'insurance premium' as a basis for future policies. Whereas current policies are based on the notion of 'no-regrets', there is a school of thought in the Netherlands which believes that the scientific information about the likelyhood of climatic disruption is sufficient to warrant going beyond 'no-regrets' and also implementing policies which carry moderate net costs as a sort of insurance premium against future risks.

GLOBAL STRATEGIES FOR THE LONGER TERM

Considering the name and purpose of this conference, in the third and final section of my paper I would like to share some thoughts about the direction that a global energy strategy should take. The question that we all face is: How can we reconcile our exploding demand for energy with the imperative of maintaining a viable, sustainable global ecosystem? In answering this question we need to look at medium as well as long term options. Our energy strategy will have to be designed to meet both the economic and ecological requirements of sustainable development. The Brundlandt Commission defined sustainable development as "development that meets the needs of the present generation without compromising the ability of future generations to meet their needs".

The economic criteria of sustainability require that energy sources be adequate to meet the basic needs of a growing world population and provide material well-being by fueling economic development and growth in all nations of the world. In order to meet the economic criteria we must ensure that the rate of depletion of fossil fuel does not exceed the rate at which alternatives are developed and applied.

The ecological requirements of sustainability go beyond the issue of climate change. They reflect the need to preserve the 'carrying capacity' of the environment, protect the biosphere and prevent more localized forms of pollution. The notion of 'carrying capacity' as a leading principle is not new. I came across it in Aldo Leopold's 1948 book, "A Sand County Almanac", one of the major American roots of an environmental ethic. It is sad to note that 45 years later the world community is still struggling to fully comprehend his conclusions. The Dutch government will embrace this concept in its future policies.

What does this mean as it applies to the issue of energy use and climate change? A mean global warming limit of 2 to 2,5 degrees Celsius, for example, would require a cumulative fossil carbon budget for the planet as a whole of 300 bilion tonnes of carbon between 1985 and 2100. According to Florentin Krause's study "Energy policy in the Greenhouse" (3) this would mean that the following milestones would have to guide a worldwide fossil phase-out: by 2005 we should be back to 1985 levels, by 2015 we should see a 20% reduction relative to 1990, by 2030 50%, and by 2050 75%. A close to optimal compromise between international equity and practical feasibility might be found by allocating the total budget of 300 billion tonnes on a 50-50 basis to industrialized countries on the one hand and developing countries on the other. Based on this allocation, industrialized nations would have to achieve a 20% reduction by 2005, a 50% reduction by 2015 and a 75% reduction by 2030. Developing countries would have to aim at stabilizing their fossil fuel consumption early in the next century and limit increases in the meantime to about 50-100% above 1985 levels. The long term level of carbon releases from developing countries would have to be about half the current rate and should be reached by the middle of the next century.

The question, of course, is how do we do this? It is no accident that the strategies of both the EC as a whole and of the Netherlands emphasize the demand side in the medium term - that is, the accent is on improving the efficiency with which we utilize our existing energy sources and on minimizing their less attractive side effects. The most environmentally friendly energy resources, as well as the most cost-effective, is energy conservation.

In the longer term, however, we will have to develop new sources of energy. Fossil fuels are not only dirty, they are also finite in supply. At current levels of demand, fossil fuels are being exhausted at a rate that is on the order of 100.000 times faster than the rate at which they are being created. Projections about remaining reserves differ, but they all agree on the fact that one day the reserves will be gone. And the relatively cleaner fuels - natural gas and oil - will be gone long before the dirtiest - coal. It will be up to us to apply our ingenuity and creativity to develop substitutes which can meet both the economic and ecological requirements of sustainable development.

Before I move to the role of developing countries in a long term global strategy, I would like to say a few words about nuclear power. Whenever the greenhouse effect is a topic of discussion the specter of nuclear power raises its head. Proponents argue that it provides a source of electricity which is free of CO_2 emissions and, therefore, that it is sustainable. While I would have to concede that nuclear power may be an attractive option from the narrow point of view of CO_2 abatement, I have to point out that this fact by itself does not mean that it is a sustainable energy alternative. The ecological criteria for sustainability which I used earlier go beyond protection of the biosphere and also include preservation of the environment's 'carrying capacity' and prevention of localized forms of pollution. When we see the size of the area made uninhabitable by the accident at Chernobyl I think we have to conclude that, until the day that someone develops intrinsically risk-free nuclear power - and that means not only intrinsically risk-free in operation but also without unacceptable risks due to long lived radioactive waste - we cannot include it among the sustainable energy sources. In the energy/carbon tax proposed by the Commission of the EC there is also a tax on nuclear power. This reflects the Commission's opinion that carbon-free does not necessarily mean sustainable.

I will conclude with the developing countries. As I noted earlier with respect to carbon dioxide emissions, a real imbalance between industrialized and developing countries also exists with regard to energy consumption. The industrialized countries of the world currently account for the lion's share of energy consumption. With less than 30% of the world's population, they consume almost 75% of the world's energy. But this imbalance in per capita energy use between the industrialized and the developing countries cannot be expected to continue. The developing countries in the Third World and the formerly planned economies of Eastern Europe aspire to, and have every right to, the same standard of living that the highly industrialized nations enjoy. If current trends continue, the developing countries will consume as much fossil fuel in about 20 years as the industrialized countries do now. This would more than double their emissions of fuel-dependent CO2, which presently account for roughly 19% of total CO2 emissions worldwide. So even dramatic improvements in energy efficiency will be insufficient if they are confined to the industrialized economies of Western Europe, North America and the Pacific Rim. The challenge, than, will be to see to it that current trends are **not** allowed to continue while at the same time ensuring that the populations of developing countries enjoy the basic amenities to which they have a right. I am confident that this is a challenge we can meet if we want to.

REFERENCES

1. Commission of the European Communities, Directorate-General for Economic and Financial Affairs,"The Economics of Limiting CO_2 Emissions",in:European Economy, Special Edition 1, Bruxelles/ Luxemburg 1992.
2. Ministry of Housing, Physical Planning and the Environment, the Netherlands, "Measures taken within The Netherlands Programme on Climate Change," CCD Paper 3, Leidschendam 1992.
3. F. Krause, Energy Policy in the Greenhouse, Berkeley, 1990.

DISCUSSION

MS. BLEVISS: I'd like to encourage questions at this point. In particular, I would like to encourage questions that have an element of interplay between the representatives of developed and developing countries up here. Any questions?

How does Canada propose to reduce CO2? Does Canada's plan and/or experience provide any lessons for Eastern Europe or Russia which is also energy intensive?

MR. YANG: Canada plans to develop a management plan first. In Canada, the provinces own the natural resources, so energy policy is basically a provincial jurisdiction. For example, in Alberta, where most of our oil is found, they control the wellhead price. Each provincial jurisdiction is going to go after what everybody here has called no regrets, which the environmentalists claim is a backwards way of approaching the problem, because you're committing to take only those actions which make sense if global warming doesn't occur, which is somehow avoiding the problem.

But certainly, energy efficiency. The movement in terms of electrical generation is more problematic, because much of Canada's generation already is hydraulic. Ontario has a particular problem, because it has a serious over capacity and a large nuclear dependence. There isn't any big opportunity in Canada in the electrical generation sector to reduce CO2.

The next level is cars and transportation. I would say there would be probably some movement there. Our fuel economy standards generally lag the U.S.'s by two or three years, so that gap could probably be made up. In terms of businesses and buildings, our building code standards are not uniform across the country; once again, they're provincial jurisdiction. In the construction industry, it's unlikely there's going to be very much gain there in the next five or six years.

Lessons for Eastern Europe? I can't think of any directly. Our steel industry, which has done quite a flip over the last five to ten years in terms of its energy efficiency improvements, is being squeezed out of the world's market for other reasons, not related to energy efficiency. I would guess that the formerly Eastern bloc countries likely have a very serious problem with lack of reliable electrical services as well as capital. We wouldn't be able to claim that kind of capital shortage in Canada. It's really institutional problems for us rather than just dire survival.

MS. BLEVISS: I have a question for the members representing the developing countries. That is, inviting comment on the discussions from two different perspectives, both on the potential for tradeable credits in terms of the implementation reality in the developing world and also the proposal about splitting the available fossil fuels between the north and the south. So I would invite Mr. Mathur or Mr. Kalinin.

MR. MATHUR: From a policy point of view, tradable permits are probably one of the best methods for reallocating resources. However, I foresee political problems. The political problems would deal with initial allocations. Would it be on a per capita basis, would it be on a political basis? Would colder countries get extra share because of the need for heating, or would the warmer countries get an extra share because of the need for air conditioning?

The second point has to do with how efficiently the tradeable market would function. As you realize, efficient market requires two things. One is ease of entry and exit, and the second is information. Developing countries sometimes feel that in such a market they would be squeezed out because of these problems. You realize, of course, there are a very few developed countries and a very large number of developing countries, which implies that there can be a formation of a cartel. So the prices that get into the market may not be fair market prices.

I don't think trading permits will come out that soon. The first place that trading permits would come out is between the industrialized countries, where there is much greater equity. Therefore, I really don't see the developing countries and the developed countries all coming into the same boat. I think the first thing that will happen is, within the European Community we will see some sort of mechanism coming out. This was referred to with Greece, Spain, and Portugal. They will probably increase their energy. How would the commission go about doing this?

There is a bit of learning to be done. I think the trading permit system will occur in the EC. It would then occur in the developed countries and then probably in the world as a whole.

As far as the second question is concerned, how do you split fossil fuel reserves between different countries? That becomes a question of national sovereignty. If a country has a large amount of oil or coal, and you say that half of this belongs to somebody else, they won't like it. I am reasonably sure that the U.S. is supposed to have one eighth of the total coal reserves in the world. The fact that they don't use all of them is largely because they have access to other sources. But even the U.S. is not going to say that we are willing to barter our longterm options at this point of time.

This is something which will be important especially if global warming actually happens, which I think people will be more clear about in 2010 or so, when the statistical trends become more apparent. I think we will start talking seriously about barter. I don't have an answer now, except to say that I don't see it being politically possible.

MR. KALININ: In 1988, the former Soviet Union occupied a strong position in the oil world market, especially in Europe. The big production of oil in 1988 was about 615 million tons a year. That is the largest in the world. Since then, for three years the production of oil in the former Soviet Union dropped to the level of about 400 million tons only. That is the expectation for this year, 1992.

The result of this catastrophic drop could not be explained by any natural reasons, but by the chaotic disorganization of the economy of the country. The near perspective is the disappearance of Russia and the Soviet Union from the global oil market, even more gloomy, more sober, in ten or 15 years the post-Soviet economy will change from an oil exporting economy to an oil importing economy.

In this context, if the world economy is interested in keeping the oil market in stable condition, one of the keys to is to keep afloat the oil branch of the Russian economy. That is impossible now without large Western investments in the oil branch. So despite the huge oil reserves, in Russia and some other countries of the former Soviet Union, the key to those treasures is in Western hands. Russia has not the capital to exploit those reserves, and if the Western investments do not come on time to help the Russian oil branch, it will mean that there would be no reserve to split between Russia and the other countries. I would stress again that Russia would join the company of oil debtors in this case.

MR. STURM: Very briefly, I would just say that Dr. Mathur's comments are very eloquent and reflect the way I see the situation. However, he also brings a negative perspective to it in saying these mechanisms are not going to be the mechanisms to move this agenda ahead. Dr. Yang referred to the no-regrets approach, simply doing everything that's cost effective outside of the perspective of global warming. Let's not overlook that the global warming problem might be enough to put in place the institutional changes necessary to allow markets to implement no-regret policies and mechanisms.

MS. BLEVISS: I have one question for Mr. van Zijst I think it's uppermost on peoples' minds. How do you see any repositioning, if at all, by the European Community as a result of the change in American administration?

MR. VAN ZIJST: A lot of people including people who I work with here in the United States and back home in The Netherlands ask me almost eagerly, so now things are going to change? One of my initial reactions is, that is to be seen. One thing which is already clear in these first couple of weeks since the election is that things are changing in the way Washington is breathing. But I think the economic agenda of Mr. Clinton will be up front very dominantly. Any other topic for a long time will be very silent. I hope for this country that the health issue will come up, because I think that's one of the major things this country has to cope with. Then there will be room for some other things, and among them is environment. With Vice President Gore in the White House, of course there will be room for a lot of other things which relate to both the use of new technologies and environmental techniques.

So the overall expectations are quite modest. Let's see how these first hundred days go before we come to any final conclusions on the environmental friendliness of this new administration.

Getting that situation back to the tax, I don't see Mr. Clinton at this moment being in favor of a CO2 tax or any other form of tax. The conditionality in the European Community's proposal, says that the rest of the OECD countries do not do something which is similar to the effect the CO2 energy tax would have for the European Community. In that case, the European Community's proposal would not go through. That doesn't mean that American has to come up with a tax. They can come up with other types of methods which would aggregate at the same level of success as a tax would do in the European Community. But again, first see how the first months of the new administration go. The odds are good, but we are a bit reluctant to applaud it beforehand.

INTEGRATING CONCERNS OF DEVELOPING AND DEVELOPED NATIONS INTRODUCTION

Carol Werner

Environmental & Energy Study Institute

In this session, the topic is integrating the concerns of developing and developed nations. We are acutely aware of how complex the whole issue of global climate change is and also how critical it is that we understand the differences amongst our various countries, societies and communities. We really need to be flexible and innovative in terms of how we approach this. There isn't one silver bullet, one magic solution.

Indeed, I think that there are probably a whole panoply of solutions and alternatives that we need to look at. We have heard a lot about the promise of technology development and what it can do in terms of moving us down a path of more sustainable energy policy, and how we can start to approach a carbon constrained world .

We have also heard about the barriers, and about ways in which those can be overcome. I think that it is clear that the issue is very complex, which makes it both frustrating and also very fascinating when we understand how interrelated everything is.

U.S. leadership is critical if we are to move forward.

It is also very clear that while there are a lot of uncertainties and problem areas confronting us as we deal with global energy strategies, at the same time, we probably all realize that things will only become more difficult the longer we wait.

We have heard many illustrations about policies that make sense for a whole variety of reasons. We really need to think about what our overall goals are and what are all the ways in which we can achieve them.

We need to listen to each other. We need to learn from each other. We need to think about technology and information transfer in terms of not just north to south, but also south to south and south to north, because we need to develop an understanding of what makes sense, what is appropriate in different places.

We have very interesting papers to follow that will help us illuminate how to think about integrating the concerns coming from the developed world and from developing countries.

So, I would like for you to have those in mind as you look over the past several days suggestions and proposals and strategies for dealing with these problems and also, at the same time, have in mind what I guess we can call the climate puzzle, all of the different interests of political institutions, cultural and other social conditions represented by the myriad of nations signing the Rio Convention, the framework agreement on climate, different stages of economic development of those societies, these differences in costs, the impacts that change

Global Enery Strategies: Living with Restricted Greenhouse Gas Emissions, Edited by J.C. White, Plenum Press, New York, 1994

itself will have differentially on societies and nations around the world, different intensities of transportation use and fuels, the impacts of forest management policies, which have been quite contentious over these last several years and, of course, the distribution of fossil fuel producers in producing states around the world.

This is the backdrop for those specific criteria or evaluative questions I was asking earlier. What I will pose for you is what I see to be a set of several linked strategies, which have been under development for a number of years now, which Carol described as emissions trading. This is a little description in pictorial terms of the framework developed for the acid rain emissions trading strategy under Title IV of the Clean Air Act, which EDF had quite a bit to do with.

It involves the simple, basic premise of the government in its usual inimitable way in dealing with pollutants of any sort, issuing permits, permits whether determining the restrictions in terms of limits and rates of emission or whether permits of the sort used in trading, describing an allowance to emit in terms of tons, are usually the initial starting point for government regulation.

After that, though, each one of these sources -- and we know, of course, that sources are of different size, have different possibilities in terms of costs of dealing with or managing pollutants, like sulfur dioxide, each one of which will have by next year a continuous emissions monitor mounted in its stack. That monitoring will include annual reports on carbon dioxide, which not only improves our understanding and database associated with emissions and developing the emissions inventory in the United States, but at the same time provides us with and provides emitters with a conscious and tangible record of what they are doing, much similar to the reporting under the toxic inventory data system caused and focused people on rather immediate and direct management of those emissions.

That tracking then is in terms of emissions over time, is what the government focuses on. That is the general strategy. One of the important things about the Acid Rain Emission Treaty approach, which I think has been really underappreciated by many in the climate change arena, is the fact that while legislatively, we manage emissions on a pollutant by pollutant basis, the reality is that when you touch off a base load, goal-fired plant or a gas turbine or any one of these other fossil fuel-based generating stations, we are committing ourselves to a vector of emissions.

We are going to get sulfur dioxide in some amount. We are going to get nitrogen oxides. We are going to have carbon dioxide, et cetera. These are not individual pollutants produced by one single process.

When we look at the -- and this is an EPA projection done for the evaluation of alternative policies under acid rain. We translated those projections into carbon dioxide terms. EPA for the first time in the evaluation of the Clean Air Act did both high and low projections -- they did a high and a low scenario. One of the major differences between this high and low case in terms of evolution of carbon dioxide emissions from the utilities sector had to do with the extent to which we are going to have business as usual over time.

And of course because these were scenarios, there were differences in the underlying projections associated with things like the penetration of cogeneration. How much demand side management or other conservation was going to be utilized to damp down this future trajectory associated with economic growth and the performance of the U.S. economy in terms of CO2 emissions.

Because this strategy is agnostic with respect to control technology, it means that if you have an effective demand side management program and that translates into reduced SO2 up the stack, you get for the first time without intervention of utility regulators or other special government subsidies or incentives, you get a direct financial and economic reward for that conservation investment in the form of avoided SO2, which you can then sell into the market place.

So, automatically, you have created not only an incentive for conservation, but a reason as to why this high emissions scenario can, in fact, if aggressively pursued on its own merits, be deflected down to this lower emission trajectory.

If you look at the intersection of SO2 control policies, which at the same time reduce CO2, you can get a very aggressive reduction in the amount of carbon dioxide, which we actually ultimately will have to deal with or somehow compensate for by emphasizing things like the switches to natural gas. Natural gas, I am sure, as you have heard, for years now gives you a hit with respect to conversions from either coal or coal fire with coal up to 30 percent, plus the conversion of oil to gas.

So, that is one element of the acid rain implications for CO2 management. The second element in that acid rain strategy, which also, I think, is underappreciated, is the fact that while we think about ourselves as being linked by free trade over long distances, we don't often think about the fact that our -- and we think about our atmosphere as being linked over long distances -- we don't often think about our energy systems being integrated or linked and the reality is that if you look at Canada and the United States, both of those systems are gridded and locked together all the way down into the heartland of the U.S.

I am sure that David, from his days in that area, can probably tell us something about the practical realities of decisions made in Ontario in implications in places like Tennessee, although not necessarily Sacramento.

What that means practically is that winter peak time purchases made by Ontario in the United States mean that the Canadians are simultaneously importing not only our power, but also the sulfur dioxide associated with its generation. So, we have the opportunity before us now with not only the acid rain program in the United States, but with the review that is going on in Canada for its acid rain strategy and the determination of how it is going to actually meet and implement its longstanding objectives, the opportunity to create companion markets for the environmental commodities in the form of sulfur dioxide that is this trans-boundary pollutant, in addition to the normal market, trans-boundary market, that we have right now, in the commodity of energy.

EPA now for a little over a year has been studying with the provincial government of Ontario what our two economies and energy systems would look like if they were integrated or linked with the greenhouse gas or carbon dioxide base on a trading system; another natural adjunct.

So, I have described three main building blocks then of a comprehensive integrated trading type strategy that exists already in the Clean Air Act of 1990. That is the CO2 monitoring and reporting. This monitoring and integration of the Canadian Acid Rain Program for the important international precedence of trans-boundary exchanges and the flexibility of emissions trading in rewarding CO2 reducing SO2 control strategies.

The next domestic piece of this puzzle in the foundation here is the National Energy Policy Act of 1992, on which the ink is barely dry and which few, if any, have started to think about implementing regulations. In that legislation, there is something called the Global Warming Title, which creates a greenhouse gas inventory, a database and a tracking mechanism for all of those companies and utilities out there, who recognizing the handwriting on the wall, have said that they wished to voluntarily reduce greenhouse gas emissions.

There will be guidelines set by the Department of Energy in consultation with EPA, concerning the appropriate crediting and tracking for things like energy efficiency improvements and buildings for EPA's green lights investments, for whatever golden carrots there may be out in the patch, for investments in forestry, for improvements in vehicle efficiency, all beyond the mandated government minimums. We can expect to see those emerge over the next 18 months or so.

We also have in the form of Rio-based commitments the foundation for an international system of transactions. This is under the somewhat unlikely language of joint implementation, which allows for cooperation between two or more nations in achieving the

Rio objectives. There has already been some very active and quite aggressive development work, led by Norway, a country which not only introduced but championed this particular strategy and proposal for introduction into the Rio agreement.

While it remains a vague concept, there are a number of experiments around the world. Some of these have been in the form of demonstration projects, which preceded the development of joint implementation. For example, applied energy services, various investments in Guatemala or in Paraguay, experiments with either reforestation or forest preservation for the point of compensating for greenhouse gas emissions.

There have been a number of exploratory investments in terms of improving efficiency in the rebuilding economies in Eastern Europe. Discussions with the Russians under an EPA-led team about reducing methane leakage from its natural gas, production and support system and also under the Forests for the Future Program proposed in Rio by the United States, there are a set of experimental investments in various international forest management options, which are being conducted by -- developed at this time by EPA.

We have, that is, EDF has, a cooperative agreement with Russia now to develop a demonstration project in the Soviet Union around the range of forest management practices, focused specifically on the question of what international and/or national institutions need to be developed, to be able to guarantee and assure the yields from such international investments.

How can we create and tap into the self-interest in those countries to monitor and maintain those investments over time? What sort of surveillance and monitoring is required? What type of insurance program should be established to guarantee the integrity of what are essentially long-lived investments.

Each one of these pieces are likely to -- or are focused on the development and to inform the development of guidelines for joint implementation, which the parties to the Rio Convention will be developing. All of this in anticipation that, in fact, the nations of the world are going to find the voluntary and open-ended commitments made in Rio unsatisfactory and will find the need to, in fact, strengthen those with a firm commitment to not just the targets, but a associated timetable.

I hope that I have pained a very short, brief and very idealized sketch of what I see to be the three main policy strands right now, the first being the Clean Air Act of 1990, the second being our own National Energy Policy Act of '92 and the third being the joint limitation provisions of the Rio Convention, all of which are under active development now; all of which have actually, in fact, generated investments, which have reduced greenhouse gases and all of which I see pointing to the solution to this problem as the creation of a companion set of markets for environmental commodities that are the natural analogues ad produce the automatic compensation for environmental damage that is normally associated with the free and untrammeled operation of a free market in normal goods.

Thanks for your time and I will be happy to answer questions.

(Applause.)

POPULATION, ENVIRONMENT AND THE IMPLICATION FOR ENERGY USAGE

Michael S. Strauss

American Association for the Advancement of Science

Expanding world populations have a potentially adverse impact on the global environment. While this largely intuitive statement has broad acceptance, articulating the precise nature of the relationship between population and environment has been difficult. There is a significant lack of understanding about this relationship and the actions needed to obtain the information necessary to developing rational and focused responses.

With regard to energy usage, many of the analyses are equally intuitive. It would seem obvious that increasing world populations will bring with them a greater demand for energy in a variety of forms. Electricity and fuels for industry and transportation, fuels, fertilizers, and chemicals for agriculture will be needed to supply the needs of society both in the industrialized nations and in those aspiring to greater benefits. We have heard enough of the facts about the disproportionate consumption of the developed nations of the world to easily conclude that as the lesser developed nations seek to achieve economic, industrial, and social gains they, too, can become disproportionate consumers.

Thus, we have little difficulty concluding that there is a direct link between population and energy consumption. Again, however, articulating this link and seeking solutions that would address our need to achieve policies and technologies of more sustainable energy usage--whether populations are increasing or declining--may itself, prove challenging. Despite this, defining the nature of the interactions between environment, consumption (including energy usage), and population is essential to achieving a sustainable society.

This paper is not intended as an examination of energy policies or technologies that could address the above challenge. Rather, it provides an overview of those dimensions of the problem associated with providing food to a growing human population. Energy is so readily and easily associated with electric lights, airplanes, automobiles, and industry that the average individual may miss recognizing agriculture and the production of global food needs as major sources of energy consumption.

THE POPULATION CHALLENGE

In 1992 the U.S. National Academy of Sciences and the Royal Society of London issued a precedent-setting joint statement.

Global Enery Strategies: Living with Restricted Greenhouse Gas Emissions, Edited by J.C. White, Plenum Press, New York, 1994

"If current predictions of population growth prove accurate and patterns of human activity on the planet remain unchanged, science and technology may not be able to prevent either irreversible degradation of the environment or continued poverty of much of the world."[1]

The United Nations Population Fund (UNFPA) had earlier underscored the tremendous challenge that growing human numbers were with regard to providing basic resources.

"As the 20th century draws to a close, the world is confronted by a daunting challenge: to bring growing human numbers and their growing needs into balance with the natural resource base that underpins much development. Choices made during the next 10 years will determine, to a large extent, the future habitability of the planet. The collision between human numbers and the resources needed to sustain them will become more acute in the remaining years of this century and beyond."[2]

The World Bank has provided projections of populations using differing assumptions of fertility for the period 1985 to 2160.[3] Their base case assumes a decline in fertility from 1.7 to 1.0 by 2030. This would result in all countries reaching replacement fertility levels by about the middle of the 21st century and a total global population of 12.5 to 13 billion individuals. Assumptions implying a more rapid decline in fertility would achieve replacement levels at about the same time, but at a somewhat lower total population of 10 billion. Such declines have been shown to be possible in, for example, Hong Kong, Jamaica, Mexico, Costa Rica, and Thailand. By contrast, slowly declining fertility--at half the rate of the base case--would mean that replacement levels would not be achieved until well into the

TABLE 1. POPULATION GROWTH RATES AND DISTRIBUTION OF WORLD POPULATION BETWEEN INDUSTRIALIZED AND DEVELOPING REGIONS, 1950-1986

	AVERAGE ANNUAL RATE				
INTERVAL	Industrialized regions (B)	Developing regions (A)	Difference (B-A)	% of world pop in developing regions	% of growth pop. in developing regions
1980-1986	0.66	1.98	-1.32	76.9	91.6
1970-1980	0.78	2.23	-1.45	74.4	89.6
1960-1970	1.04	2.41	-1.37	71.7	86.2
1950-1960	1.26	2.07	-0.81	69.9	79.8
1940-1950	0.35	1.44	-1.09	67.5	90.0
1930-1940	0.85	1.28	-0.43	66.4	77.3
1920-1930	0.91	1.11	-0.20	66.1	68.6
1900-1920	0.92	0.52	0.40	67.9	53.8
1850-1900	1.05	0.53	0.52	73.3	54.5
1800-1850	0.83	0.31	0.52	78.1	53.4
1750-1800	0.62	0.47	0.15	79.3	73.6
1650-1750	0.33	0.34	-0.01	79.3	79.4

Source: D. Gale Johnson, Can There Be Too Much Human Capital? Is There A World Population Problem?. Office for Agricultural Economics Research, Paper No. 92:10. The University of Chicago, 1992.

22 century and at a total world population in excess of 22 billion. Declines of this nature are seen in Paraguay, Sri Lanka, Suriname, and Turkey.

Since the middle of this century the disparity between developed and developing nations has become particularly acute (Table 1). The growth rate in developed nations has declined each decade since 1950 from 1.26 to 0.66 in the period from 1980-86. By contrast, the rate in developing countries has remained above 2 (2.41 in the decade 1970-1980) with the exception of a drop to 1.98 in 1980-86. More significantly, the percentage of world population growth occurring in developing countries is now more than 91 percent. It is, thus, not surprising to see that more than 75 percent of global population is found in the developing world.

These numbers, coupled with the desire on the part of countries to achieve growth in per capita income, can have significant implications for the environment, as illustrated in the World Development Report, 1992.[4] Availability of clean water and adequate sanitation improve with per capita income; and, in general, particulate air pollution declines in urban centers. Sulfur dioxide emissions initially increase with higher incomes initially, and decrease at higher levels, presumable because of an ability to afford more environmentally sensitive technology. Not surprisingly, perhaps, both municipal wastes and carbon dioxide emissions increase dramatically as income rises. When combined with the pressure of rapidly growing populations, these increases can have severe adverse environmental impacts. Please note, that in this analysis sanitation, which concerns the availability of a health safe environment for the individual, is separate from the problem of municipal wastes which, in part, must address how to deal with the byproducts of providing adequate personal sanitation.

SUPPLYING A GROWING POPULATION: THE AGRICULTURAL CHALLENGE

Garrison Wilkes of the University of Massachusetts has provided us with a very graphic view of the challenge confronting modern agriculture.[5] He points out that early in the first 20 years of the next century the number of people alive and, thus, needing daily food will be ten percent of all that have ever existed on the planet. By that time a single year's global food production will equal the amount of food produced in the century from 1850 to 1950 (because of growth patterns demand for food by developing nations will have doubled. Finally, in the first two decades of the next century the amount of food produced will equal all that has been produced since the beginnings of agriculture 8,000 to 10,000 years ago.

Clearly, we are facing a daunting challenge. Once all agriculture was for subsistence and was practiced in gardens not unlike many seen in rural areas around the world (except, of course, that the diversity in these reflects centuries of exploration and exchange). At the dawn of agriculture humans were at a density of about one person per 25 square kilometers. Today the density is 25 people per square kilometer and from 600 to 1,200 per square kilometer in urban areas.

In part this has been the result of our agricultural success. In simple ecological terms, the species has increased because its primary food source has increased. Or, in the words of the writer of the Old Testament book of Ecclesiastes, "When goods are increase, they are increased that eat them" (Ecclesiastes 5:11). Past concerns of food adequacy have been met with dramatic increases in production capacities.[6] The so-called Green Revolution where genetic alteration of wheat and rice enabled several-fold increases turned the world back from what many predicted was sure disaster. But we are again approaching a time where increases in crop production do not appear to keep pace with population growth.[7] Further, efforts such as the United Nations Conference on Environment and Development and the Agenda 21 blueprint that resulted, have raised an awareness that future solutions must consider the importance of balancing population needs with preservation of the global resource upon which we all, ultimately, depend.

So, where do we go from here? Some would argue that the road to sustainability is a road back to simpler days. That, however, is the way back to subsistence and we clearly face a challenge far beyond the capacities of those days. The only option is to move forward to develop technologies that are both highly productive and sustainable. By sustainable I mean the successful management of resources to satisfy developing human needs while not sacrificing future options, preserving or enhancing environmental quality, and conserving the natural resources upon which the global population depends. Or as the Bruntland Commission defined it, sustainable development is "meeting the needs of the present generation without compromising the needs of future generations."[8]

For agriculture the promise of improving practices in many regions suggests that there remains much that can be accomplished in providing for growing populations. As the numbers increase, however, we must not sacrifice the basic resources of air, soil, water, and biological diversity. I do not suggest that we abandon modern technology and practice, but rather that greater effort be devoted to understanding the processes that underlie agriculture as it relates to these resources, and to the development of technologies to conserve, manage, and sustainably use them.

IMPLICATIONS FOR ENERGY USAGE

It should be clear that the sustainable and efficient use of energy is woven throughout the fabric of these remarks. While this paper does not present data about the link between population and energy usage, a powerful illustration exists in an image from the National Aeronautics and Space Administration (NASA).

From the vantage point of an orbit around the globe, a satellite can provide a unique view of the earth. From there the earth at night displays a myriad of lights with particular brightness in those areas of greatest population density. It is easily possible to locate all of the world's major population centers by the light they produce at night. This "Earth at Night," as NASA calls the image is a graphic illustration of the fact that growing populations place an ever increasing burden on the world's energy resources.

FINDING SOLUTIONS

The most obvious solution to the problems of population, consumption, and the environment is to rapidly reduce population growth to the level of replacement. However, demographers warn us that there are longer term consequences of such changes in the demographic structure of the population. Policies of population growth must take into account the impact of significant changes in age distribution on both present and future societies.

Some would suggest that we are already beyond the limits of our global resources. This may be true, but perhaps the optimist's view would be more productive. That is, that we strive to overcome what we presently view as barriers to sustainability. Recall that before the Green Revolution it was generally thought that we were at or beyond the limits of global food production. However, this does not imply an endorsement of those positions that would suggest that there will always be a technological solution that will come to our rescue. Technology holds the keys to many of our future needs, but it will be limited if we do not preserve an adequate resource upon which it can be developed.

The World Bank has outlined the challenges to a sustainable society and the ways to achieve it.[9] The report suggests a need for a partnership between developed and developing nations. It calls for improved and new technologies, and increased investment in their development. It argues for reform of trade and capital markets and for environmentally responsible growth in world economies. It notes the need for increased assistance to reduce poverty, confront environmental damage, and slow population growth. And it proposes the participation of industrialized nations in the financing of environmental initiatives in developing nations, especially those that have globally significant consequences, such as ozone depletion or warming due to carbon dioxide emissions.

For population growth the answers are not as simple as might appear. Its is sometimes convenient to see population growth as the central element forcing all other issues--control population and you control all other problems. In fact, it is a significant factor, but many other things have contributed to global and regional problems and must be addressed as well. The World Bank report argues that controlling population will require progress on four fronts:

1. Raising incomes of poor households.
2. Reducing child and infant mortality
3. Expanding education and employment opportunities, especially for women.
4. Increasing access to family planning (This author would stress that this should be voluntary family planning).

Note that the above list, rather than simply addressing fertility, aims to address the fundamental issues that underlie both the growth in populations and the development of sustainable societies. The list can--and should--be made much longer. The key for us is to recognize that achieving a sustainable society and a stable population is an ambitious, but achievable goal.

FOOTNOTES

[1] U.S. National Academy of Sciences and the Royal Society of London. 1992. Population Growth, Resource Consumption, and a Sustainable World. Joint Statement 27 February, 1992.

[2] United Nations Population Fund (UNFPA). 1991. *Population, Resources and the Environment: The Critical Challenges*. UNFPA, New York.

[3] World Bank. 1992. *The World Development Report, 1992*. The World Bank, Washington, D.C.

[4] World Bank. 1992. *The World Development Report, 1992*. The World Bank, Washington, D.C.

[5] Wilkes, Garrison. 1992. *Strategies for Sustaining Crop Germplasm Preservation, Enhancement, and Use*. Issues in Agriculture, No. 5. Consultative Group for International Agricultural Research, Washington, D.C.

[6] Swaminathan, M.S. 1993 Genetic Enhancement of Sustainability of Yield. (In press). *In*, D. R. Buxton et al. (eds.) International Crop Science I. Crop Science Society of America, Madison, WI.

[7] Swaminathan, M.S. 1993 Genetic Enhancement of Sustainability of Yield. (In press). *In*, D. R. Buxton et al. (eds.) International Crop Science I. Crop Science Society of America, Madison, WI.

Ehrlich, P.R. and A. H. Ehrlich. 1990. *The Population Explosion*, Simon and Schuster, New York.

[8] World Commission on Environment and Development. 1987. *Our Common Future*. Oxford University Press, New York.

[9] World Bank. 1992. *The World Development Report, 1992*. The World Bank, Washington, D.C.

GREENHOUSE GASES AND EMISSIONS TRADING

Alice LeBlanc

and

Daniel J. Dudek

Environmental Defense Fund

Atmospheric concentrations of carbon dioxide and other greenhouse gases have grown rapidly since the beginning of this century. Unless emissions are controlled, the world could face rapid climate changes, including an increase in temperature and shifts in weather patterns.

Major sources of carbon-dioxide emissions, the main contributor to climate change, are the burning of fossil fuels and deforestation. Other greenhouse gases include methane (natural gas) and nitrous oxides. Fossil-fuel burning and landfills cause methane to escape into the atmosphere. Methane also is produced by cows and sheep and in rice cultivation. Nitrous oxide comes from fertilizers and burning of wood and fossil fuels.

Any attempt to control the greenhouse effect should include all major contributors, not just carbon dioxide.

Global cooperation is essential. According to a 1987 study by the World Resources Institute, the three largest emitters of greenhouse gases contributed 40 percent of worldwide emissions. The top 10 countries contributed 65 percent.[1] As many countries as possible should agree to greenhouse-gas emissions targets, however, to make sure that overall emissions are limited in the future and to ensure that reductions are made in an efficient and cost-effective way that will impose the least burden on the world economy.

As a first step, nations must agree on an overall global limit based on the best scientific knowledge of what is required to protect the Earth's natural systems. The Montreal Protocol sets a precedent for limiting global greenhouse-gas emissions. Under this international accord, more than 70 nations have agreed to consumption limits on the most damaging chlorofluoro-carbons, and production of these chemicals will be phased out by 1996.

The next step is to divide the global greenhouse-gas limit among nations. Overall limits and allocation rules must take into account environmental goals, equity, and economic efficiency. The allocation could be based on historical emissions, the size of the economy, the population, or some combination of these factors.

Reprinted from Forum for Applied Research and Public Policy, Vol. 8, No. 2, pp. 40-44 summer 1993.

An effective agreement also will have to establish reporting terms and enforcement provisions. Any country that is part of the agreement would need to have in place a reliable way to certify that limits are met. Because a country might have an incentive to underestimate its emissions, an international agency should have the authority and resources to assure reliable national reporting.

Emissions can be reduced by increasing the efficiency of industrial processes and electricity generation, producing more efficient appliances, producing more efficient automobiles, switching to a less carbon-intensive mix of fuels, or developing alternate fuel sources such as biomass, solar, or wind.

Because trees remove carbon dioxide from the atmosphere and store it as carbon, another way to compensate for emissions is through forest management, forest preservation, and tree planting. Capture and use of methane from coalbeds or landfills, prevention of methane leakage from pipelines, and changes in fertilizers, rice cultivation, and livestock feed are options for limiting other greenhouse gas emissions.

CONVENTION FOR CHANGE

The Intergovernmental Panel on Climate Change (IPCC), a group established by the United Nations Environment Programme and the World Meteorological Organization, acknowledged the threat and set negotiations in motion for a climate convention. At the United Nations' conference on Environment and Development held in Brazil in June 1992, most of the nations of the world signed a Framework Convention on Climate Change as a first step toward a treaty to limit greenhouse-gas emissions.[2]

Under the terms of the convention, developed countries will aim to return to 1990 emission levels of carbon dioxide and other greenhouse gases not controlled by the Montreal Protocol.

Principles for meeting obligations include cost-effectiveness and consideration of all relevant sources and sinks. Emissions inventories and national reports will be prepared periodically by signatory countries.

Through joint implementation, the framework convention provides the structure to develop a market-based approach. Joint implementation means that developed countries may make investments for greenhouse-gas reductions in developing countries and receive credit toward their obligations.

Benefits of this approach are a flow of funds to the south for environmentally beneficial projects, cost-effective reductions for the north in meeting obligations, and a vested interest on the part of the party providing the funding that results are achieved.

Projects eligible for joint implementation are likely to include any project that reduces greenhouse-gas emissions, such as increasing energy, appliance, or automobile efficiency; switching to less carbon-intensive fuels, renewables, or biomass; increasing efficiency of manufacturing processes; reducing methane emissions from gas pipelines, coal beds, or landfills; and forestry projects that result in either increased sequestration of carbon dioxide or decreased emissions from deforestation.

EMISSIONS TRADING

Estimates of how much it will cost to reduce carbon-dioxide emissions vary widely from country to country. In the United States, the range is from $1.50 to $65 per metric ton, depending on the method used.[3]

Table 1 lists the costs of some offset options in the United States. The estimated average cost of planting trees is $9.50 per metric ton of carbon removed; the cost for conservation measures is estimated at $6.50; and switching from oil to natural gas at $5. The cost of installing and operating scrubbers at electric generating plants to remove carbon dioxide before it enters the atmosphere is $65.00.

These estimates assume removal of 130 million metric tons of carbon dioxide will be required. In actuality, greater reductions will be necessary. For example, if only a 20-percent reduction were required, the United States would need to reduce approximately 900 million metric tons from today's levels and even more in the future.

TABLE 1
Carbon Dioxide Offset Costs

Option	Cost per Metric Ton
Fuel Switching	$5
Conservation	$6.50
Shade Trees	$1.50 - 7.50
Scrubbing	$65
Tree Planting	$7 - 12

Source: Daniel J. Dudek and Alice LeBlanc, "Offsetting New CO_2 Emissions: A First Rational Greenhouse Policy Step," Contemporary Policy Issues VIII (July 1990).

International emissions trading is a way to lower costs and expand reduction options for the benefit of all. As part of the implementation process established in a climate convention, it would work as follows: an international agency would assign each participating country allowances, stated in tons of carbon dioxide, and equal to its allowed emissions. The sum of the allowances would equal the global greenhouse gas limits. Emissions limits would be monitored and enforced by procedures established in the climate protocol. These monitoring and enforcement procedures would be required with or without trading.

A country would be free to buy and sell allowances, but its emissions in a given year must not exceed the allowances it holds. By emitting less than the allowed amount, a country would accumulate allowances to sell.

Trading will occur when there are differences in the costs of reduction between two countries. This provides incentive for the country with the lower-cost option to reduce emissions below its limit and sell those excess reductions at a profit to a nation whose reduction or offset costs are higher. The nation with the higher reduction costs, on the other hand, would have an incentive to buy excess emissions reductions rather than reduce or offset in the more costly way.

As long as the overall carbon-dioxide reduction limit is maintained, the environment does not suffer from this transaction. However, both countries gain from the trade.

The trading system is voluntary. Trading occurs only if both parties gain. If a country decides not to trade it still must meet the quota established by the climate convention.

TRADING EXAMPLES

In addition to cutting costs, emissions trading provides an effective means of technology transfer. Former Soviet bloc countries provide ample opportunities for technological improvement that will result in greenhouse-gas emissions reductions at lower cost than might be achieved in the United States. In Poland, for example, heat is, in many cases, provided by coal-fired "local" boilers that are old, unregulated, inefficient, and large emitters of other pollutants such as nitrous oxides and sulfur dioxide. A U.S. firm might retrofit these boilers or replace them with a central heating source, thereby reducing multiple pollutants.

Partial payment for the new equipment would take the form of carbon-dioxide allowances. Both parties gain by this transaction. Carbon-dioxide emissions are reduced, Poland obtains more efficient equipment at a reduced cost, and the United States gets credit for some of the emissions reductions at a lower cost.

Legislation was introduced in Congress in 1991 that would require major new stationary sources of carbon dioxide to offset their emissions and gain carbon-dioxide offset credits. The legislation would have allowed offsets to be created in several ways: improvements in appliance and automobile efficiency, fuel switching, increased efficiency in electricity generation, tree planting, preservation of old-growth forests, and capture of coal-bed methane. The proposed legislation also included the potential for credits for forestry projects in other countries. Offsets involving methane would be converted to carbon-dioxide equivalent credits.

The National Energy Policy Act of 1992 contains a provision to document and record voluntary greenhouse-gas emissions reductions. This provision is an abbreviated and voluntary version of the stationary source carbon-dioxide offset legislation introduced in 1991. It holds the potential for future regulatory credit for early greenhouse-gas reductions, including international and forestry offsets. Some forward-looking U.S. utilities have initiated programs to voluntarily reduce greenhouse-gas emissions, including investments in international forestry projects.

The acid rain portion of the 1990 Clean Air Act sets an example for large-scale emissions trading. It limits the amount of sulfur dioxide that electric utilities emit and creates a market for sulfur-dioxide emission credits. A utility that reduces emissions below the limits set by law can sell emission credits to another utility.

A global market for carbon dioxide emission credits can offer an efficient means of achieving the emission limits established in an international agreement. The benefit of this efficiency to the world economy is enormous. Billions of dollars can be saved by allowing the country that can achieve reductions at the lowest cost to sell its excess reductions to others. In an increasingly integrated world economy this cost savings translates into increases in human welfare everywhere.

ROLE OF FORESTS

Forests store large amounts of carbon. The current rate of destruction of tropical forests has been estimated to contribute 20 percent of anthropogenic global carbon-dioxide emissions. The potential also exists to increase carbon sequestration and reduce emissions through reforestation and better forest management. If an international agreement were reached that included obligations in all countries, some countries that contain forests might have to reduce emissions or might wish to create additional emissions reductions below the limits set by the convention. That could be done by limiting the destruction of their forests and planting trees on degraded or barren land.

If forest acres can be converted into carbon-dioxide equivalents, such countries might find it economically attractive, as well as environmentally sound, to slow the destruction of their forests beyond what is required by the protocol or to plant trees to take additional carbon out of the atmosphere. By pursuing this strategy they would earn greenhouse - gas credits to offset emissions from industry or agriculture or to sell to foreign firms or governments.

A trading program authorized under an international convention could provide substantial economic incentives to protect forest beyond what is required by the international agreement.

How might a system of greenhouse-gas trading work to foster forest preservation? If a country with tropical forests chooses to meet its obligations under a climate convention or to generate extra allowances by reducing deforestation, it must make sure no more than an allowed number of acres is deforested. It could give or sell limited permits for deforestation to landowners. These permits could be converted into carbon-dioxide equivalents.

If monitoring indicates that less than the allowed amount of deforestation has occurred, excess allowances may be available to sell to other countries or to save for future years. By the same token, if the permits are worth more than the yield from deforestation landowners who receive permits might decide not to deforest but to sell their permits for cash to buyers requiring extra carbon dioxide allowances. The condition of permit sales would be that the land remain undisturbed.

Worldwide greenhouse-gas emission limits will become increasingly difficult to achieve over time as pressures for growth and development intensify. Preservation and sink enhancement of forests alone will not solve the problem. Many other sources of reductions will need to be exploited. However, including forestry carbon-dioxide credits in a global emissions-trading system will provide an additional source of credits and generate revenues for preserving tropical forests.

CONCLUSION

Compared to other policies, greenhouse-gas allowance trading among nations has several advantages.

It would use market forces to achieve a cost-effective choice of options for emissions reductions.

It would encourage innovation in reducing emissions through efficiency, new technology, and by capturing and storing them in biomass.

The incentive for forest protection is an important feature of emissions trading. Allowing additional permits to be created through tree planting and forest preservation lessens the impact of greenhouse-gas control on the economy. It also creates opportunities for those nations that have forests as part of their natural resource wealth. Trading also provides a way of allocating scarce resources, namely carbon-dioxide permits, over time and space.

Lastly, a system of marketable permits provides a relatively simple means of determining performance. The sum of permits transferred to control authorities at any one point in time measures total emissions.

International trading in greenhouse gases provides a tool to help achieve global reductions in the most efficient way. Allowing the least-cost supplier of emissions reductions to supply larger amounts lessens the cost for all. By including forest preservation and sink enhancement as a reduction strategy, a value for the forest is created that protects biodiversity and human habitat as well.

Joint implementation under the climate convention is the first step toward an emissions trading system in greenhouse gases. Several joint implementation projects are in the planning stages, including projects to generate international forestry offsets in Russia by U.S. utilities.

Reducing greenhouse gases to slow global warming will require tremendous resources. Bringing a policy to this endeavor that minimizes costs and encourages innovation is critically important for the survival of the natural environment as we know it.

NOTES

1. World Resources 1990-1991, a report by The World Resources Institute in collaboration with the United Nations Development Programma and the United Nations Environment Programme (NY: Oxford University Press, 1990).
2. J.T. Houghton, et al., eds., Climate Change, the IPCC Scientific Assessment (NY: Cambridge University Press, 1990).
3. Daniel J. Dudek and Alice M. LeBlanc, "Offsetting New CO_2 Emissions: A First Rational Greenhouse Policy Step," Contemporary Policy Issues 8 (July 1990).

TECHNOLOGY TRANSFER

Robert E. Cole

Global Climate Coalition

Working for the aluminum industry, I am very much aware of how important environmental concerns are to the public and to the advantages which can come to companies which produce products which are environmentally compatible. The recyclability of aluminum is one of the advantages which our industry proudly advertises. Not only does recycling reduce the amount of waste which goes into landfills, but recycling saves about 95 percent of the energy needed to make new aluminum from its original ore. The largest use of aluminum in the United States is for aluminum beverage cans. the recycling rate for aluminum cans exceeds 60 percent. The largest growth market for aluminum in the future is projected to be the automotive industry, where aluminum's light weight is an advantage to automakers in their efforts to increase fuel efficiency and emissions performance without compromising safety, overall performance, or size. Recycling is important in the automotive market too. Today, over 85 percent of aluminum automotive scrap is reclaimed and recycled.

Kaiser has benefited substantially from its environmentally compatible products. We have also benefitted from our environmentally related technology transfer business here in the United States and abroad. Technology sales is an important line of business for us. An example of that technology transfer business is in the Commonwealth of Independent States (C.I.S. - formerly the Soviet Union) where we are engaged in projects that will help the Russian aluminum industry improve the energy efficiency and emissions performance of aluminum smelters. These technology transfer arrangements are good business for us and for the Russians as well as good environmental stewardship. Our technology transfer activities with the C.I.S. are not limited to one way transfers. The Russian aluminum industry has made technological advances which could generate important technology transfer business for them, too. In fact, we believe that opportunities exist for combining U.S. and Russian technological capabilities to achieve new technology developments and create future technology transfer opportunities for companies in the two countries and in other countries around the world.

I have cited these examples of environmental technology transfer not just to give you a commercial for the fine company for which I work, but more importantly to help make the point that a lot of environmental technology transfer is currently occurring within the United States and around the world. The technology transfer is occurring within the private sector in response to peoples' needs as reflected in the marketplace, often with little or no government role.

As a former history teacher, I am a little amazed at how much in vogue technology transfer is among politicians, academicians, and many of the rest of us today. The image of technology has come a long way since the early 1800s when the "Luddites" of England rioted and smashed the machines that they feared would dehumanize the workplace and take away jobs. Of course, we know now - almost 200 years later - that the Luddites greatly overestimated the potential applicability of technology in eliminating human work and greatly underestimated the benefits of technology in improving living standards. We also know that the Luddites were not entirely wrong. Not all technological developments of the last two hundred years have been beneficial in all ways and at all times. Still, public discussions today generally reflect the view that in both economic and environmental areas progress can be best achieved with more and better technology, rather than less.

The international Framework Convention on Climate Change reflects this view. During the negotiations of the Framework Convention, the INC stated that the developing countries requested that the "best available, environmentally sound technologies" be transferred to them on a "most favorable basis" and with the developed countries providing "adequate and additional financial resources" to assure the transfer occurs expeditiously. Technology cooperation was a major issue throughout the negotiation and the final Convention devotes considerable attention to facilitating such cooperation.

Technology transfer has also been a major focus of the Global Climate Coalition, a broad-based organization of business trade associations and companies representing virtually all elements of U.S. industry including the energy-producing and energy-consuming sectors. The Global Climate Coalition believes that the United States can make important contributions to improving the global environment and improving conditions for economic development by encouraging technology transfer to developing nations, Eastern Europe, the C.I.S., and other developed nations.

For the past two decades, the U.S. has been at the forefront among the nations of the world on environmental policy, technology development, and implementation. In 1990 the American People and industry spent more than $115 billion on protecting the environment. Our leadership position, with our private and public research and development capabilities, gives us the edge in providing technology to countries whose economies are in transition, including Eastern Europe and the C.I.S., as those countries focus on the dual objectives of environmental improvement and economic development.

The private sector's participation in environmental technology transfer is of critical importance. In fact, technology transfer is principally a private sector to private sector, rather than government to government activity. The worldwide movement away from centrally planned economies toward private markets further underscores this fact. U.S. government technology transfer policies should recognize and facilitate private sector transfer.

U.S. Government policy must also recognize that a strong and growing economy and a robust industrial sector are prerequisites for addressing domestic and international environmental challenges. With an expanding economy, American industry can continue to develop and produce technologies that will make the American economy more efficient, and through technology cooperation make it possible for developing nations to expand their economies in an environmentally sound manner, while benefitting American companies at home.

The Global Climate Coalition has an active program to address national and international issues and to assist U.S. Government activities related to technology cooperation. The Coalition has participated in numerous policy meetings with representatives of Congress and Federal agencies such as the Environmental Protection Agency, Department of Commerce, Department of State, Department of Energy and the Council on Environmental Quality; addressed technology cooperation issues before the INC and other international forums; cosponsored a conference with the U.S. Department of Commerce on technology transfer to Eastern Europe; and assisted in the development of the U.S. proposals for a

Technology Cooperation Corps. We are currently preparing a Resource Guide to the technology cooperation programs of the U.S. Government.

Even more significantly, U.S. industry has an extensive program of environmental technology programs overseas. Many companies that are members of the Coalition have joint ventures or other mechanisms in which they engage in technology cooperation projects in developing countries.

Private sector efforts would be greatly assisted by better access to information on opportunities for technology transfer business and government technology assistance programs and a coordinated U.S. technology would improve our competitive position in this arena. The Environmental Protection Agency, Department of Commerce, and Department of Energy all have important resources and programs which could provide needed information to industry. Those programs should be coordinated, and private sector interaction should be improved.

Importantly, inadequate protection of intellectual property rights creates substantial barriers to technology transfer. Without guaranteed protection for patents, trademarks and copyrights, U.S. companies have a strong disincentive to pursue the costly work of technological and industrial innovation and to transfer that technology overseas. To facilitate technology transfer, U.S. must continue to demand that foreign governments and firms protect U.S. technology, and property rights.

A final barrier to technology transfer is the uncertainty related to the antitrust implications of private firm collaboration. Private firms frequently are reluctant to establish joint research and technology development programs with other firms. Yet, because of the prohibitive costs and highly speculative nature of technology development programs, private companies are hesitant to sustain the costs alone. Thus, rapid development and diffusion of technology suffers. By reducing the uncertainties related to antitrust enforcement, private companies could form joint ventures, merge their resources, and develop and introduce new products more quickly.

The Federal Government could assist U.S. technology transfer activities by:

1) Helping countries prepare accurate and detailed needs assessments.

2) Providing additional analyses and information on environmental technology needs and market opportunities to U.S. business through the embassy and consulate staff as well as through the Commerce Department's International Trade Administration staffs.

3) Identifying and eliminating impediments to technology transfer.

4) Facilitating the entry and acceptance of new technologies where appropriate.

5) Promoting U.S. businesses as sources of environmental technology to meet the needs to developing countries and their industries.

6) Supporting research, development, demonstration and commercialization programs.

The U.S. Government should also work with international financing and economic development agencies to provide information on U.S. environmental technology capabilities and assist U.S. technology suppliers to meet the environmental needs identified by other countries.

The provisions of the Convention encourage information exchange and the transfer of technology - hard and soft - from the developed world to those countries that are now developing their economies and their resources. If properly prepared, the National Action

Plans being developed by each country that signed the Convention can provide a wealth of information about the capabilities of countries to supply technology and the needs of the countries that wish to receive technology. The Global Climate Coalition has recommended that a section of each plan be devoted to a "Technology" component to draw together information each country needs if cooperative efforts among the suppliers and the receivers of technology are to be advanced in the most expeditious manner.

Specifically, the Global Climate Coalition has recommended that the National Plans include specific discussions of the topics particular to the individual countries seeking technologies or countries that will supply technology and resources.

Following is an outline of items which should be included:

I. Technology Assessments

A. Conditions within the country that would be conducive to improvement by technology
   1. Specific needs of the Public Sector
   2. Specific needs of the Private Sector

B. Specific technological requirements or technologies available
   1. for direct reduction of emissions
   2. For improvements in energy efficiency
   3. For enhancement of sinks
   4. For adaptation or mitigation requirements

C. System support requirements or system support available
   1. Worker training
   2. Management training
   3. Management systems
   4. Maintenance and repair systems
   5. Financing requirements

D. Changes in existing technologies which may be required to meet country needs

E. Barriers to technology transfer
   1. Identification of legal and institutional barriers
   2. Remedial actions required to remove barriers

F. Sources of technology and systems
   1. Indigenous capability to supply
   2. Technology import requirements
   3. Research capability to assess and adapt existing technologies to unique country requirements

II. Country Technology Acquisition and Utilization Plan

A. Priority programs for technology applications
   1. In the Public Sector
   2. In the Private Sector
      a. Industry
      b. Final consumers

B. Acquisition or export programs
   1. Purchases
   2. Joint ventures
   3. Foreign aid requirements

C. Description of the support programs of the receiving/supplying country

D. Maintenance and monitoring programs planned

A combination of the above items, with the material that is included in the U.S. developed outline of a National Action Plan including 1) the statement of national circumstances, 2) future greenhouse gas emissions trends, 3) adaptation actions, and 4) national mitigation actions, will enable the private sectors of each country to work together with respective governments to identify the opportunities that best combine the capabilities and resources of the technology-supplying country with the needs of the technology-receiving country.

The international Framework Convention on Climate Change presents a significant opportunity for new levels of international cooperation on transfer of technologies related to energy and environment. To be successful, the cooperation must be founded on sound economic principles and government technology transfer policies must recognize and facilitate private sector transfers. The Global Climate Coalition believes that with a strong program of technology cooperation, the United States can assist developing nations and those economies in transition to expand their economies in an environmentally sound manner.

ENERGY STRATEGY -- IS A COMPREHENSIVE APPROACH POSSIBLE?

S. David Freeman

Sacramento Municipal Utility District

One of the key elements in this whole global climate change and energy dilemma is to recognize all the choices that we have and perhaps exercise some of the better ones before it is too late.

I want to underscore what my good friend, Dr. Dudek, said earlier. In all the ideas that we consider, let's put a little more emphasis on what is politically feasible. We have a lot of ideas that keep getting recycled. They come up over and over again, and one of the reasons they keep getting recycled is that they have been rejected again and again as politically infeasible.

We sent a sulfur tax up on the Hill in the early seventies and couldn't get a Senator to introduce it. I think that the admonition that we heard earlier to examine all our ideas, not on the basis of how intellectually stimulating they might be in the rarified atmosphere that we have here, but whether they can be implemented in a national policy, is very good advice.

It is clear to me that energy efficiency is absolutely essential as a foundation for dealing with the global climate problem. Consider the number of people who are already on earth and the number of people that are going to be here, unless we develop the renewable technologies they are going to be burning coal or oil or natural gas or, even worse, trying to build and operate nuclear power plants. We need to face the cold, hard truth that while efficiency is absolutely essential, it is insufficient. We have had $4 dollar a gallon gasoline in Europe for a long, long time, but I haven't seen any electric cars over there.

The renewable and new technologies are not automatically developed by higher energy prices. It is trite and overly simplistic, but on a discounted cash flow basis, the world isn't worth saving.

We just are not going to get private enterprise to make the long term investments in anything as fundamental as photovoltaic cells, and solar power plants and other things, without some industrial policy and well-defined and aggressive involvement by the government, because the benefits are for society and the benefits can't be captured sufficiently by any particular company to make those kinds of investments feasible. The payout is too long.

Now, those are hard facts that just have to be recognized if we are going to deal with the problem.

I read an article by a very bright man, George Will, not long ago, in which he said that in a Gallup Poll of scientists only 52 percent thought that the global warming problem

Global Enery Strategies: Living with Restricted Greenhouse Gas Emissions, Edited by J.C. White, Plenum Press, New York, 1994

was serious and, therefore, why is Vice President Gore and all these people creating such a fuss about it.

Well, now stop and think a second. If you thought there was a 52 percent chance you were going to be hit over the head and mugged when you walked outside, you would probably invest a little bit in some protection, a bodyguard, a gun or something. It is because we have never experienced global warming that we are so divorced from nature and we have the arrogance of our ignorance. If it hasn't happened, then we are not going to deal with it.

That is the society that we live in and that is what we are up against politically. The abnormity of the challenge has to be recognized. A smaller scale example is that no one had ever experienced the kind of winds that they had in Hurricane Andrew. So, the damage was far greater than anyone predicted because the models couldn't visualize it. It was what the nuclear people say, an incredible accident. They said that before Three Mile Island.

If it is something that we haven't visualized as possibly happening, then we assume it is not going to happen. That is a terrible obstacle for dealing with this problem. So, the important thing is to try to do the things that are politically feasible and fortunately, the solutions that will do the most good are not the hardest. And we are doing them in Sacramento, California, which may seem like Disney Land to someone from the East Coast, but it is really not.

Sacramento is really more like a Midwestern city. It is an ordinary American town. We have got about a million people in the county. It is the only place on earth that the people by vote decided to shut down an operating nuclear plant. They did that before I got there.

Basically, the utility is unilaterally disarmed, if you want to think of it that way. The plant was half of our generating capacity and it was just shut down. So that we have gone to, in the interim, replacing it by purchasing power from our friendly neighbors, PG&E and the Edison Company, and in the meantime, we are putting together a program for self-sufficiency.

This is a little, old utility. We don't have the kind of money to get on the nightly news, but we are going to reduce the carbon content of our electric power supply by 20 percent in this century and it really isn't all that hard. We are replacing Rancho Seco with some highly efficient natural gas-fired cogeneration plants, a wind project and some hydropower from the Northwest.

The carbon content is reduced primarily because we are getting the power now from gas-fired plants with heat rates of around 11,000 and replacing them with new cogeneration plants that will have heat rates of around 7,000. But we also are meeting all of our growth in the capital of California, all of it, a hundred percent, with energy efficiency.

This is not just a PR stunt or a day dream. This is a program that has been ongoing now for the better part of three years and we have kept our peak load, our managed peak, flatter than a pancake. You know, it is not something that you just dream about. We push a button on a hot day in the summer and we do a notch test and 175 megawatts goes off the system just like that through our load management program that we have worked up over the years. We call it the Peak Core. We give people $20 a month off in the summertime and they let us interrupt the air conditioning on the hottest days of the year and it works out fine.

But that is not efficiency. That is just load management. We do know the difference. We have a very comprehensive set of energy efficiency programs and it is not gadget oriented. We just take the avoided costs and we don't even include the environmental externalities because we don't need to. The efficiency option is low cost enough that at roughly speaking, about 4 cents a kilowatt hour, we are buying the electricity that the people in Sacramento would otherwise be wasting.

We give them rebates to get super efficient refrigerators and trade in their old ones that use twice as much electricity per unit of cooling. We destroy those old refrigerators and recycle the metal and send the bad stuff back to DuPont or whoever makes it.

We plant trees at the proper places around people's homes and we look at a tree as an air conditioner and we have got the numbers that show that ten years from now, those trees will be virtually the equivalent of central air conditioning in Sacramento.

And lighting, of course, is one of the major items in our program. We are investing over 6 percent of our revenues each year into this energy efficiency program, a higher percentage of our revenues than any utility in the country. Our rates are not as high as some back in the East, so that may not be the fairest test of all, but we certainly have got a very intensive energy efficiency program and our business plan for the rest of the decade will show that it is a steady program that will keep our managed peak load essentially flat on into the next century.

The energy usage will go up slightly, but not sufficiently to cause another peak. Basically, we will be building the equivalent of a six or a seven hundred megawatt power plant, perhaps even eight hundred megawatts. We are in the beautiful position that if we overconserve, we just build less renewable energy resources in the future. We are committed to completing the replacement of Rancho Seco in '96 with a block of another 400 megawatts of either renewables or very advanced technologies, such as fuel cells. Either that or if we overconserve, it will be that much less.

We have a truly integrated resource plan in which the winners are our customers. We intend to keep the real price of electricity essentially flat in Sacramento and, indeed, the people that participate in our efficiency program will probably see their bills go down in real terms, although we don't know how much inflation there is going to be.

This program is not a hard program. It is hard in the sense of managing and implementing, but in terms of our customers, it is the most popular thing that SMUD has ever done. We are actually helping people reduce their bills and people can understand that.

The other part of it is that the money is being spent largely in Sacramento County and, with the recession, the 45, or 50 million dollars a year that we are investing in efficiency is keeping a lot of small businesses and contractors in business that wouldn't otherwise be. So, it is politically attractive and actually attractive to the community. Those are important things to keep in mind.

It is the one option that can truly be a win-win situation. Now, SMUD is in an easier position to implement this because we are not regulated by a PEC and our stockholders are our customers. As a publicly-owned power system, we do not have the triangular problem of the stockholders, the consumer and the utility. We just have a straight relationship between the consumer, who is in a sense the stockholder, and the utility. That makes it, somewhat easier.

On the supply side of our situation, we are working with Southern California Edison in a project to make a power tower, with molten salt as a storage mechanism to get some dispatchable thermal solar power. That experiment is underway and we expect toward the end of the century to have reasonably economic solar power.

We are putting photo voltaic cells on the roofs of a number of homes just to get experience and we expect that PV power, especially located in the remoter parts of our service area, will cut down on the number of feeders we need to build. This gives us the situation where PVs can be compared, not with the bus bar cost of electricity, but with basically the retail cost and also eliminate some capital investment. With that kind of thinking, we see the PVs coming in in the latter part of this decade and being a significant player.

We are installing fuel cells. We see the fuel cell as a very important piece of equipment to generate electricity in large and small quantities in the future.

The perhaps most interesting thing that SMUD is doing -- and I guess this way of putting it is a bit much, but looking down the road, we are thinking about putting Exxon out of business in Sacramento County. I mean, not literally, that we are trying to take over the entire transportation market.

We see electricity as replacing petroleum in the transportation market. We see it not necessarily just as a selfish way to improve our load factor by charging batteries at night. I don't know any other way that the renewable resources or even our domestic coal or nuclear power is going to displace imported oil, other than in the form of electricity.

The idea of the fuel cell powered by hydrogen is an even better idea, but then the hydrogen has to be created with some sort of electrical supply, by electrolysis of water. So fundamentally we are looking at electric cars. We have a partnership with the regional transit agency. We are building a trolley bus system in downtown Sacramento; back to the future.

We are looking at jitney services with electric vehicles so that people that live in the suburbs and have to take trips will have an electric vehicle. We are working with the Air Resources Board in California, and with our local air quality group and SMUD has become a partner for clean air.

We have transformed not just the image but the performance of this utility from one as a polluter to one who has a major role to play in developing clean, zero emission vehicles. These really are the only answer to the local air pollution problems and, of course, the only answer over time to the build up of the gases that are probably causing global warming on a scale that we can't tolerate.

We have popular support for these programs in Sacramento and Sacramento is not a radical community. It is a fairly conservative community. We have gone to the people with these programs. There has been a lot of public participation. We have a publicly elected board of directors, who have to be responsive.

We had elections this fall of two of the directors. Everybody ran on the ticket of supporting the SMUD program. Don't let people tell you that moving to efficiency in renewables is pie in the sky and not feasible. It is very feasible. What has to happen in this country is that we need to stop just talking about it and start doing it.

DISCUSSION

MS. WERNER: You have now heard a whole variety of examples of ways that we can approach energy policy solutions and how some of these things can be done innovatively to meet multiple objectives, and that it can be pretty exciting to watch it happen.

The first question is for Dan Dudek. Could you talk a little bit more in terms of why is impact analysis in terms of social, environmental, economic effects not being adequately funded or pursued.

MR. DUDEK: I think a lot of the question of impact analysis has to do with how we make environmental policy. If you look at the history of most environmental legislation, you find some initial concern voiced either by advocates or more often by scientists, about the potential posed by some pollutant or environmental danger.

That is usually followed by a period in which the affected interests on both sides of the issue or on three sides; the industry, the government and environmental advocates all conduct model wars at varying degrees of intensity.

Usually the modeling wars relate to the political attention given to the environmental issue. In the case of acid rain, this went on for about a dozen years with models both of the cost and impact, as well as the funding of research to resolve fundamental underlying scientific uncertainties.

The reality was that the political consensus and will had not yet really been developed to solve the problem. No one had applied a very common sense test to continuing to research the problem. Is additional funding and research reducing or providing enough information to change our fundamental understanding of the problem?

In other words, would we do anything different if we had the results from this new research expenditure? If we wouldn't, then why are we making it? The answer usually is that we are buying time for the political process.

It is a sort of horse and cart question, which really relates to the intensity of the political focus and scrutiny and the priority associated with the problem.

MS. WERNER: A couple of questions for Mr. Cole. The first one is at the Rio Summit, more than 400 European and Japanese companies presented technologies at the Earth Tech Exhibition; yet, only 25 American companies were present. What accounts for this small number of U.S. companies particularly relative to the major economic competition that U.S. companies would seek? And what should be done about that?

MR. COLE: I can't speak for a lot of companies. I can speak for my own company. We were putting our resources into the actual technology transfer programs we had underway in Russia; into looking at new technology resource opportunities in specific countries in South America, like Venezuela, in Africa, like Mozambique, and in other areas of the world. We made a judgment as to where was the right place to use our limited resources. We were going after real projects as opposed to putting our money into demonstrations down in Rio.

MS. WERNER: A question for David Freeman. In your resource planning, are you taking climate change productions into account for air conditioning loads, cooling water availability, hydro availability, changes in wind regimes or cloudiness for solar cells, et cetera?

MR. FREEMAN: Well, our air conditioning program, if I can be a little facetious about it, is our climate change program. People couldn't live in 112 degree heat in Sacramento and work as well without air conditioning. We have been through six years of drought. It was raining when I left. So that we have a huge variation in what our hydro potential will be and we obviously have to take that into consideration.

Whether we have gotten sophisticated enough to get into the other items that you mentioned? I doubt, quite frankly. We feel like the most important thing we could do is to reduce the carbon content of our power supply and help develop the technologies to take that relatively clean electricity and use it to substitute for petroleum in the transportation sector.

Those are our main thrusts. The points that you made, I don't think we have done much with. I will say that we have done some very interesting research, though, on the impact of coloration on the roofs of homes in the urban heat island impact. We are finding that if we can paint all the roofs white and get white pigment into all the asphalt, that there is maybe as much as a ten degree differential that could be achieved. We are finding that putting these trees up, just next to the compressors is having a marvelous impact.

When you start fooling around with innovative programs, you learn things that you really hadn't even thought you were going to learn. That is not answering the question, I know, but that is the best I can do.

MS. WERNER: Another question for Robert Cole. Do you believe that the emissions trading concept talked about by Dan Dudek would facilitate private sector technology transfer and if so, do you support the implementation of such a trading scheme?

MR. COLE: I think whether or not the emissions trading concept would facilitate transfer depends upon how well it was designed and how it would fit into the other programs. Right now, I am not in support of emissions trading. I think that you have to make subjective judgments in order to make emissions trading work.

The first judgment you have to make is what is the cap? Without a cap on emissions, there is no value to the trade.

The second judgment is are you really accomplishing your overall systems objectives by focusing on pricing one element of environmental concerns differently from others?

There are ways that we can increase or decrease our emissions in our plant by consuming more electricity and having the utility be the one that incurs the emissions. Emissions trading schemes don't always work through all the economics of those things in such a way that it gets you the optimum result.

It seems to me that in a new system, there is a burden of proof on the proponents to show that it would achieve the environmental results that we would like to achieve.

MR. DUDEK: First of all, what you heard is the standard position of those who are just saying "no." What you are witnessing here is the standard dialogue between an environmentalist and an industry representative about does she or doesn't she.

The cap is not needed. We have all across the country a set of emissions trading protocols which have been in operation since the mid-seventies, based on the idea of an offset that new sources entering into the systems, or major modifications associated with increases in emissions have to purchase and buy compensating reductions in the emissions market. There is no cap on that system. There is a regulatory burden, an obligation placed upon those contributing to new emissions in the area as a way of facilitating continued economic growth.

The idea that you require some agreement on a global cap for emissions trading to occur is a red herring designed to scare the bejeezus out of everybody who hears it.

MR. COLE: By the way, I did not say "global." I think if you are locating a plant in a specific location and you have to buy offsets, there is a cap in that location.

MR. DUDEK: Again, I beg to differ because in fact most of the systems in Title I non-attainment areas are based on regulations which control the rate of emissions and not the total load of emissions. Consequently, they are uncapped systems.

David Freeman was complaining about the problem of the differential regulatory treatment of stationary sources versus mobile sources in the United States. The reality is that the mobile source contributions, the non-attainment problems in the U.S., are essentially uncontrolled. Nobody tells people how much to drive their cars and it is VMT, vehicle miles traveled, which is wagging the dog's tail in that system. This is causing political pressure on the easy, politically vulnerable, stationary sources who have to face up to increasing control costs at their vulnerable, highly visible smokestacks.

So, the system is uncontrolled. It is not capped. The fig leaf provided for how we generate reasonable further progress while allowing economic growth to occur, is the idea that increases in emissions from industrial development are compensated through the offset policy. That is the way non-attainment and Title I of the Clean Air Act functions in this country.

There are real economic values associated with offset transactions without a capped system. There has been quite a bit of economic analysis in support of that, and a number of companies, taken these voluntary steps, invested in offset transactions.

Secondly, in terms of technology transfer, the whole point to a trading and/or offset type strategy is to reward those who would make investments in technologies overseas, that might make a difference in terms of emissions. Frequently, we hear the argument made that when we go to Eastern Europe or Russia, these people are strapped with hard currency needs. They can't afford this state of the art. They can't afford the most efficient, least emitting technologies because their real constraint is capital.

One of the ways to lessen that constraint is to reduce the cost of those newer, more capital intensive alternatives by rewarding those who would deploy those technologies overseas with credits, with allowances, which are transferable, and have a real economic value.

With respect to the question on cycle of emissions, this is clearly a question as to who is in and who is out of the system. We get back to the political feasibility issues. In the development of the Acid Rain Program in the United States, so that we did not open up World War III, industrial sources were omitted from the system. The EPA has in place in the legislation an implicit cap of about a million tons on industrial sources with an associated 8.9 million ton cap on SO2 for utility sources.

All the utility sources are monitored and controlled and regulated. We know what all of the SO2 burden is and all of the discharges from all those facilities. That is going to wash out through the investments and through the actions, through the rates faced by folks like Kaiser or Alcoa or other heavy intensive electricity users.

You may reduce your emissions at your site, but that is going to turn up in increased demands for rate hikes for industrial customers at PUCs associated with the increased burden at utilities to meet that extra demand.

So, there is no free lunch in these systems. You can't off-load power and expect to escape the responsibility, in either environmental or financial terms, for the burden caused by shifting those emissions.

MR. COLE: I might say, by the way, that Kaiser may very well be a beneficiary of the SO2 emissions trading scheme. We are looking at that. A lot depends on the base period as to where you measure improvements, but my comments on concerns about whether or not the system works were not reflective of a corporate policy that is based on us expecting not to benefit. We expect we will probably benefit. We still don't know that it is worth implementing.

MS. WERNER: Let me ask you another question, Mr. Freeman about SMUD activities. Could you provide just a little bit more detail about the wind project that you have got underway and also if you could, we have got a question about comparing the prospects for the solar power tower system that you were talking about?

MR. FREEMAN: I will answer the second, but first, it is going to be a whole lot more economic. When you pull the pure solar part out of the system and push away the gas, which you could get from your power system anyhow, you had 14 cent a kilowatt hour electricity. These solar power towers, which will be solar dispatchable in the sense that, with the molten salt, we will be able to store the solar power and have electricity on through into the evening for our full peak period; certainly the numbers suggest they will be single digit in terms of costs and provide us with real capacity benefits.

They are getting down within the range of being, "economic" in the conventional sense that that word is used without including environmental impacts.

As far as our wind project, we are very proud of that. It is a 50 megawatt project. U.S. Wind is offering us their new variable speed turbines, which are fundamentally different than all these windmills that are up there now, because they will operate at a wide range of wind speeds. With existing wind technology, most of the time the windmills are not operating because they operate only in a very small range of the wind speed.

U.S. Wind is taking all the risks. We will pay only for the kilowatt hours that they produce. We were going to have to pay 5 1/2 cents a kilowatt hour, which is somewhat more than gas-fired capacity and then the Congress, bless its lovely soul, came through in the Energy Bill, and there is a 1 1/2 cent a kilowatt hour production incentive for a project like this, if it is funded.

We have high hopes that it will be funded. And we stand to end up with 4 cent electricity, which is highly economic and SMUD is not taking the risk. We bought the land in fee simple because we figured out that what we would pay in easements would be essentially the cost of our money invested in the land and the land values are pretty stable. And at the end of 20 or 30 years, there will be better windmills and if we didn't own the site, we would have no equity in the project at all.

So, we are the owners of the site. U.S. Wind is an independent power producer. They are going to build and operate the plant for us and we have an option for another 50 megawatts. We are going to do 5 megawatts just as a test next year and we have every reason to think it will work.

MS. WERNER: Another question for Robert Cole. You emphasized that we should leave technology transfer to the marketplace, to the private sector, but at the same time present a long list of demands for the U.S. Government. Is it not more logical to leave technology transfer to public/private partnerships in bilateral or multilateral agreements than to exclusive private enterprise?

MR. COLE: I did not mean to say and did not want to leave the impression that technology transfer is solely an element of private enterprise. I believe that private enterprise will be and has been to date the fundamental provider of technology, training and systems.

I did try to identify a number of roles for government. Notice, I didn't talk a lot about subsidies in the government role, but there are a lot of things right now that are barriers. There are barriers in terms of lack of knowledge about where those markets are and government can certainly help in identifying them.

There are barriers in terms of educational levels in the receiving countries and there is certainly a role for government in improving that. I think these are partnerships, although they may not be in the traditional concept of contractual partnerships.

MS. WERNER: Actually, there was another question here, too, about Kaiser Aluminum and so much aluminum being recycled. Since recycling is so energy and cost efficient, why does Kaiser Aluminum consistently oppose statewide and national bottle bills?

MR. COLE: That is an issue we debate internally all the time, too. Our information in the industry is that a statewide bottle bill tends to depress sales of beverages for about two years and after that, the aluminum industry benefits because we take over market share from our competitors.

The difficulties have to do with, whether or not the system is working properly. Right now, we pay for metal, and the reason we have a high recycling value rate is because there is a high value of the metal. That metal price that we pay has kept a billion dollar a year industry going in the recycling area. In many communities, it pays for the disposal or recycling of glass, plastics, paper and things like that.

We are reluctant to drop what is working and producing high recycling rates in order to go to another system that may or may not work.

Depending on how the program works, you get into all kinds of side impacts. In California, its attempt to implement the program has resulted in some cases of more waste disposal because they getting back products that they couldn't do anything with. In some cases, businesses that were in the recycling area went out of business because of the way they divided up the state, allocated money, and things like that.

We would say that though there may be some value in recycling in those programs, be careful not to destroy what works; we have said basically not "hell, no," but don't destroy what is working.

MS. WERNER: To Michael Strauss. Could you discuss the U.S. position over the last 12 years; with regard to family planning internationally in terms of the U.S./Mexico City position, opposition to funding for the U.N. Population Fund and refusal to let the World Bank encourage family planning and birth control? Where do you think that policy should go? What do you anticipate?

MR. STRAUSS: Let me start by saying that is an area that I am just beginning to examine. It is certainly an area where a lot of the demographers are looking to see some significant changes.

There has been some real controversy over family planning in the U.S. position over the last 12 years. I think there have been some significant steps forward. Nancy Carter at the State Department is someone who has worked on these issues for a number of years and she is fond of saying that rather than look at what has not been accomplished over the last 12 years, we should be looking at the significant moves that people in the Population Office in State have been able to accomplish in getting the agendas moved forward.

I think one of the most significant issues in the last several years has been the activities of the Department of State and other people working to negotiate some recognition of the population question at the Rio conference. The word "population" was sort of outlawed at Rio and it was not largely a U.S. initiative to keep it out. It was a move by developing countries and by the Vatican for religious reasons; developing countries because they perceived that the population issue would derail their concerns about development. They wanted to avoid the focus of the issue turning towards population when they perceived development to be a major concern.

There was work quietly in the background to forge some compromises and at least get some of that language in. I certainly see some changes coming, but at this point, I would have to say that I am not familiar enough with the population area to discuss that in any detail.

MS. WERNER: I feel very much encouraged that we will see a shift in U.S. position on this front, but it is also really important that the whole thing be looked at again in an integrated comprehensive way. It is critical that we look at overall health care, the role of women in societies overall. There can't be a concentration solely on family planning. Otherwise, nothing works and we are looking at much deeper, more complex societal problems.

Here is a question that was bound to come up for Mr. Dudek and Mr. Freeman. Should an emissions reduction strategy consider the potential contribution of increased electrification, especially electricity from nuclear power plants, since they release no greenhouse gases?

MR. FREEMAN: I have always been impressed that the nuclear industry, especially here in Washington, has been able to perfect low expectations to an art form. We haven't had an order from anybody that has to pay for one of these suckers in over 12 years. I don't know what all this talk about nuclear power is. The plants that are out there are being shut down. We have shut ours down. The Trojan Plant is being shut down. The Southern California Edison is closing a unit down.

The units now have operating costs where you can build a brand new cogeneration gas-fired power plant and pay for the whole thing for about as much as the operating cost of a nuclear plant. I think it is time for this country to face up to a fact that the light water reactors that we have today have been a gigantic financial failure. We would have never started the nuclear power option if we didn't think we were going to go to a breeder reactor and have a very large amount of power.

All the uranium in the world doesn't give you more than a half a century's supply of energy with these technologies and they are too expensive to use. The idea of giving emissions credits to perpetuate something that has caused a Chernobyl and a Three Mile Island is not attractive. No one knows really what the impact of the radiation they are producing is; we are a generation of guinea pigs.

My own view is I don't think we ought to arbitrarily shut down the existing plants. I think the marketplace is really working. It is not Ralph Nader or Jane Fonda that stopped the nuclear option, it is the financial vice presidents of the utility industry, the marketplace. So, I think it is an almost obscene discussion to be talking about credits for nuclear power plants.

MR. DUDEK: Well, I certainly concur with everything that David said. My remarks today focused on creating a framework and focusing on the integrity of that framework. I am essentially agnostic with respect to the choice of solution or control strategy. That doesn't mean, though, that that is going to change the fundamental underlying economics. It does not mean that we ought to ignore the problems of subsidies of these strategies or that we should ignore what I see to be the three fundamental challenges faced by the nuclear industry in environmental terms.

That is, first, the inherent safety issue, toward which I think some modest progress has been made, not necessarily with respect to light water reactors. Second, long term disposal and, third, the problem of nuclear proliferation. Rather than worrying about whether they qualify for greenhouse credits or not, I suggest that there are many more fundamental underlying problems with the nuclear industry that it ought to be addressing.

MS. WERNER: Another question for you, Mr. Dudek. Previously a speaker from India expressed concern that OECD nations might form a cartel if CO2 trading was to be performed internationally. How can developing nations be assured that OECD nations wouldn't control such a market for their own benefit?

MR. DUDEK: That is an interesting point of view. The reality, though, in the development of international atmospheric agreements, like the Montreal Protocol and like Rio, is that a founding principle in the United Nations is that developing countries have a special status and somehow ought to be treated differently.

The practical translation of that principle is that you have to do something to get them into the playpen, into the sandbox in the first place. What is their inducement to participate or cooperate in some sort of global agreement? We have to address and create the self-interest for those nations to participate in the first place. There have been various strategies proposed, analyzed and considered - with respect to how developing nations can be given extra allocations, extra time, and additional allowances, - various incentives to cooperate in some sort of greenhouse gas limitation - not necessarily signing up and subscribing to a full-blown trading system.

After all, trading is voluntary and countries who have individual targets and obligations are not necessarily compelled to trade. You look internally within your individual nation as to what your opportunities and potential are.

There is no compulsion to trade and I think there is very little danger of an OPEC-like cartel, because after all, the opportunities for emission reductions are much greater than we currently recognize.

MR. FREEMAN: I just want to add this thought. OPEC has been so terribly effective, that the State Department has been going over there trying to get them to jack up the price of oil all through the eighties. So, it is hardly an example of something to fear.

MR. COLE: I talked earlier about the trading system with subjective judgments. This is part of the area. If, in fact, you make judgments that you are going to let the developing world off because they have very legitimate needs for economic growth, for refrigerators and things that are related to living standards, then you get into a real political problem here in the United States.

I am sure you all remember as part of the 1992 Presidential debates, Mr. Clinton accused AID of funding the export of a plant overseas. A trading system that favors development outside the United States will be criticized politically by the American public, not just by industry, because we may be building the facility overseas.

It was very clear in the 1992 election that the people were very worried about the economic impact of programs. Any system if it were to be implemented worldwide would have to struggle with that. Are the subjective judgments going to be such that you favor location of facilities and operations in the United States? In this case the developing world is going to get hurt; or you are going to let the developing world off in terms of compliance with these things, in which case, the political systems in the developed world have real problems.

MS. WERNER: Also, Mr. Dudek, could you comment briefly about what do you anticipate in terms of CO_2 reductions and over what time period from the three things that you talked about in terms of Clean Air Act implementation, the Energy Bill that was just enacted and the Joint Implementation Plan in the Climate Convention?

MR. DUDEK: I would have to say, I don't know. What we have done with respect to the SO_2/CO_2 interface in utility emissions is to analyze the area of overlap and say this is what the potential is. I don't know how much in the way of incentives will be generated for people to actually take advantage of them. We have state regulators and state legislatures, which are saying, no, our high sulfur coal mining industry is more important. Though shalt use scrubbers.

Scrubbers add to the inefficiency of individual plants and increase the CO_2 burden to the atmosphere. Those are local, state-based decisions that are completely legal under the framework of the Clean Air Act, yet they go in the opposite direction. So, from an analyst's standpoint, it is virtually impossible to predict what those results will be. It creates the opportunity to reward people who would take advantage of multiple pollution reductions; something that we have not encouraged, despite the fact that we are talking about an energy mix that does simultaneously produce a whole suite of pollutants.

MS. WERNER: I would like very much to thank our panel. I think we all got a lot of food for thought; ideas and good information. I hope we go away armed with renewed enthusiasm to work in important policy areas, searching for better solutions to global energy strategies. This should be an exiting year and may everybody work very hard to help make it a success.

PROGRAM

December 8, 1992

SESSION I: THE POTENTIAL FOR NON-FOLLIL FUEL ENERGY SOURCES
Chair - **Richard Ottinger**, Pace University Law School

Solar Power
Ann Polansky, Solar Energy Industries Association

Wind Technology After a Decade of Development
Randall Swisher, American Wind Energy Association

The Potential for Bio-Mass Energy and the Implications for Climate Change
David Rinebolt, U.S. Congress Staff

Hydropower
Richard Wilson, Harvard University

SESSION II: THE POTENTIAL FOR REDUCING CARBON EMISSIONS THROUGH EFFICIENCY
Chair - **Christopher Flavin**, Worldwatch Institute

The Transportation Sector
Steven Plotkin, U.S. Congress, Office of Technology Assessment

The Potential for Reducing Carbon Emissions Through Improved Efficiency in Industrial Processes
Marc Ross, University of Michigan

Increasing Economic Growth & Reducing Carbon Emissions Through Improved Energy Efficiency
Arthur Rosenfeld, Lawrence Berkeley Laboratory

Long Term Options for Energy Supply & Demand Side Management
Thomas Morron, Edison Electric Institute

Fossil Fuels & Greenshouse CO_2 Mitigation Technologies
Meyer Steinberg, Brookhaven National Laboratory

WEDNESDAY, DECEMBER 9, 1992

SESSION III: U.S. ENERGY POLICIES AND STRATEGIES
Chair - **J. Christopher Bernabo**, Science & Policy Associates, Inc.

The U.S. Energy Strategy
Edward Williams, U.S. Department of Energy

Integrating Energy and the Environment
Alex Cristofaro, U.S. Environmental Protection Agency, Office of Policy and Planning

Reducing U.S. CO_2 Emissions & the Value of Flexibility in Timing
Richard Richels, Electric Power Research Institute

The Alternative Energy Future
Bruce Henning, American Gas Association

The State Perspective
Charles Imbrecht, California Energy Commision

Why do the Strategies Differ?
David Montgomery, Charles River Associates

SESSION IV: GLOBAL ENERGY STRATEGIES: THE PERSPECTIVES OF DEVELOPING COUNTRIES AND ECONOMIES IN TRANSITION
Chair - (Sessions IV and V): **Deborah Bleviss**, International Institute for Energy Conservation

Russia/Eastern Europe
Alexander Kalinin, Monterey Institute of International Studies

Energy Use in India: Implications of Constrained Greenhouse Gas Emissions
Ajay Mathaur, Tata Energy and Resource Institute

South America
Russel Sturm, International Institute for Energy Conservation

SESSION V: GLOBAL ENERGY STRATEGIES: THE PERSPECTIVES OF DEVELOPED COUNTRIES

Reducing Energy-Related Greenhouse CO_2 Emissions: A Canadian Perspective
Bunli Yang, Ontario Ministry of the Energy

Strategies for Improving Energy Efficiency & Reducing CO_2 Emissions in the European Community & the Netherlands
Hans Van Zijst, Royal Netherlands Embassy

THURSDAY, DECEMBER 10, 1992

SESSION VI: INTEGRATING CONCERNS OF DEVELOPING AND DEVELOPED NATIONS

Opening Remarks
Carol Werner, Environmental and Energy Study Institute

Population, Environment & the Implications for Energy Usage
Michael Strauss, American Association for the Advancement of Science

Greenhouse Gases & Emissions Trading
Alice LeBlanc and Daniel J. Dudek, Environmental Defense Fund

Technology Transfer
Robert E. Cole, Global Climate Coalition

Energy Strategy--Is a Comprehensive Approach Possible?
S. David Freeman, Sacramento Municipal Utility District

PARTICIPANTS

Louis Aboud, American Gas Association, Arlington, VA
Mario Aguilar, Embassy of Mexico, Washington, DC
Donald Aitken, Union of Concerned Scientists, Woodside, CA
Thomas Anding, University of Minnesota, Minneapolis, MN
Maria Arias, Florida Power & Light Co., Miami, FL
Md.SultanulArefin Bahlul, United Club, Dhaka-1000
Carole Beal, CEI, Rochester, NY
Lira Behrens, Inside Energy, Washington, DC
William Benusa, Chevron Corporation, San Francisco, CA
Nick Berenyi, Westvaco Corporation, Summerville, SC
Christopher Bernabo, Science & Policy Associates, Inc., Washington, DC
Donald Bieniewicz, Department of the Interior, Washington, DC
Deborah Bleviss, International Institute for Energy Cons., Washington, DC
Rebecca Blood, American Public Power Association, Washington, DC
Susan Boyd, Concern, Inc., Washington, DC
William Bross, CEI, Rochester, NY
Peter Campbell, Ontario Ministry of the Environment, Toronto, Ontario CD
Edgar Chase, Clinton, MD
Marc Chupka, Joint Economic Committee, Washington, DC
Linda Ciocci, National Hydropower Association, Washington, DC
Anita Clift, A & C Enercon, Atlanta, GA
Steve Cochran, US EPA, Washington, DC
Doug Cogan, Investor Responsibility Research Center, Plainfield, NH
Joseph Cogan, PERI, Rockville, MD
Robert E. Cole, Kaiser Aluminum, Washington, DC
Peter Condon, USDA/ERS, Washington, DC
Steve Cook, Environment & Energy Study Conference, Washington, DC
Susanna Cordts, CEI, Rochester, NY
Walter Corson, Global Tomorrow Coalition, Washington, DC
Alex Cristofaro, US EPA-Air & Energy Policy Division, Washington, DC
Marjo Curgus, Pugwash, USA, Washington, DC
Louis Davis, Union of Concerned Scientists, Decatur, GA
Marjorie Davis, Sierra Club, Decatur, GA
Christopher DeChaine, NCI Information System, Arlington, VA
Helena Dempsey, Chesapeake Center, Frederick, MD
John Dempsey, Chesapeake Center, Frederick, MD
Susan Doran, CEI, Rochester, NY
Roger C. Dower, World Resources Institute, Washington, DC
Dan Dudek, Environmental Defense Fund, New York, NY
William Elliott, DOC/NOAA/OAR/ERL/ARL, Silver Spring, MD

Karen Evens, Minnesota Power, Duluth, MN
William Fang, Edison Electric Institute, Washington, DC
Jon Findley, NCI Information System, Arlington, VA
Karen Fisher-Vanden, Battelle PNL, Washington, Dc
Christopher Flavin, Worldwatch Institute, Washington, DC
Chris Fox, Center for Global Change, College Park, MD
Allan Frank, The Solar Letter, Silver Spring, MD
S. David Freeman, Sacramento Mun. Util. Dist., Sacramento,. CA
Paula Gold, New England Electric System, Westboro, MA
Patrice Gordon, Congressional Budget Office, Washington, Dc
Fran Gotcsik, CEI, Rochester, NY
Howard Gruenspecht, U.S. Dept. of Energy, Washington, DC
Ken Haller, Global Tomorrow Coalition, Washington, Dc
Matt Hammond, Environmental Info Networks, Alexandria, VA
Jaydee Hanson, United Methodist Board, Washington, DC
Mary Harris, RCG/Hagler, Bailly, Inc., Arlington, VA
Sandi Harris, National Association of Home Builders, Washington, DC
James Hartnett, University of Illinois, Chicago, IL
Jeremy Hayden, CNIE, Washington, DC
Bruce Henning, American Gas Association, Arlington, VA
Thomas Hernandez, Tampa Electric Company, Tampa, FL
Richard Hill, University of Maine, Orono, ME
Ruth Hill, United Nations Association, Charlotte, NC
Phyllis Hinterer, INGAA, Washington, DC
David Hodas, Widener University School of Law, Wilmington, DE
Laurel Hodory, Ohio University, Columbus, OH
John Howard, New York Power Authority, Albany, NY
M. Ann Howard, CEI, Rochester, NY
Jau-Inn Huang, Columbia University, New York, NY
Paul Hughes, Enterprise Advisory Services, Inc.
Robert Hughey, U.S. Department of Energy, Oakland, CA
Charles Imbrecht, California Energy Commission, Sacramento, CA
Michael Isensee, Lutheran Office for Governmental Affairs, Washington, DC
Arsuteru Ishizuka, Nissan Research & Development, Farmington Hills, MI
John Jones, US Council for Energy Awareness, Washington, DC
Russell Jones, American Petroleum Institute, Washington, DC
John Justus, Congressional Research Services, Washington, DC
Alexander Kalinin, Monterey Institute of International Studies, Monterey, CA
Thomas Kenney, NAHB Research Center, Upper Marlboro, MD
Ronald Klattenberg, Northeast Utilities, Hartford, CT
Betsey Kuhn, USDA/ERS, Washington, DC
Pat Lane, League of Women Voters, Baltimore, MD
David Lapp, Power Line, Takoma Park, MD
Tim Larson, Pugwash, USA, Washington, DC
Steve Lerrner, Commonweal, Washington, DC
Lee Link, LEAD USA, Williamstown, MA
Michelle Martens, Northern States Power, Minneapolis, MN
Ajay Mathur, Tata Energy & Resources Institute, Arlington, VA
Kenji Matsuo, Tokyo Electric Power Co., Washington, DC
Karen McCormick, DPRA, Arlington, VA
James McGranery, Jr., Washington, DC
James McNamara, Texaco Inc., White Plains, NY

Donna Mercado, American Gas Association, Arlington, VA
John Meyers, Washington International Energy Group, Washington, DC
David Montgomery, Charles River Association, Washington, DC
Thomas Morron, Edison Electric Institute, Washington, DC
Fred Munson, Greenpeace, New York, NY
Sumie Nakayama, Center for Energy & Environmental Policy Research, Cambridge, MA
Beth Nalker, Smithsonian Institution, Washington, DC
Sara Nichols, Clean Air Council, Philadelphia, PA
Jonathan Norling, Earth Resources, Ltd, Silver Spring, MD
Craig O'Hare, Sierra Club, Tucson, AZ
James O'Hear III, Catholic University of America, Washington, DC
Richard Ottinger, Pace University School of Law, White Plains, NY
Peter Pandolfini, Hopkins Uni-Physics Lab, Laurel, MD
Nancy Parks, Sierra Club, Aaronsburg, PA
Sarah Peirce-Sandner, Eastman Kodak Company, Rochester, NY
Steve Pernick, Jr., Duquesne Light Company, Pittsburgh, PA
Patti Petesch, Overseas Development Council, Washington, DC
Susan Petronio, Sterling, VA
Gerald Phelps, NAIMA, Alexandria, VA
Elizabeth Pixley, Monroe Community College, Rochester, NY
Steven Plotkin, U.S. Congress-Office of Technology Assessment, Washington, DC
Mark Popovich, Center for Clean Air Policy, Washington, DC
Robert Pratt, Global Climate Change Digest Editor-CEI, Rochester, NY
John Quackenboss, Environmental Information Networks, Alexandria, Va
William Raup, U.S. Department of Energy, Washington, DC
David Reiner, ICF, Inc., Washington, Dc
Kenneth Richards, USDA/ERS, Washington, DC
Richard Richels, EPRI, Palo Alto, CA
Barbara Richman, Environmental Magazine, Washington, DC
David Rinebolt, Office of Rep., C. Peterson, Washington, DC
Daniel Roczniak, MBD, Inc., Washington, DC
Arthur Rosenfeld, Lawrence Berkeley Laboratory, Berkeley, CA
Marc Ross, University of Michigan, Ann Arbor, MI
Margaret Rostker, Electric Power Research Institute, Washington, DC
Sam Sadler, Oregon Department of Energy, Salem, OR
David Sander, Chevron Corp., San Francisco, CA
David Schimmelpfennig, USDA/ERS. Washington, DC
Dale Schlenker, Southern Company Services, Atlanta, GA
Hugh Schratwieser, NOAA, Washington, DC
Peter Schroeder, Michigan State University, E. Lansing, MI
Charles Sills, Barbara Gauntlett Foundation, Washington, DC
Scott Sklar, Solar Energy Industries Association, Washington, DC
Joel Smith, RCG/Hagler, Bailly, Inc., Arlington, VA
Philip Sparks, Communications, Washington, DC
Meyer Steinberg, Associated Universities, Inc., Upton, NY
Melissa Stone, Kompass Resources International, Washington, DC
Fred Stoss, Oak Ridge National Labs, Oak Ridge, TN
Michael Strauss, American Association for the Advancement of Science, Washington, DC
Sheryl Sturges, AES Corp., Arlington, VA
Russell Sturm, International Institute for Energy Conservation, WAshington, DC
Eric Summers, Science & Policy Associates, Washington, DC
Nicholas Sundt, Energy, Economics & Climate Change, Washington, DC

Randall Swisher, American Wind Energy Association, Washington, DC
Elizabeth Thorndike, CEI, Rochester, NY
Steven Traut, Exxon Company, Florham Park, NJ
Hans van Zijst, Embassy of the Netherlands, Washington, DC
Phil Voorhees, Clayton Environmental Consultants, Lexington, MA
William Wagner, CEI, Rochester, NY
Barbara Wells, National Governors' Association, Washington, DC
John Bruce Wells, The Bruce Company, Washington, DC
Carol Werner, Environmental & Energy Study Institute, Washington, DC
John Wetzel, Association of American Railroads, Crofton, MD
James White, Cornell University, Ithaca, NY
Richard Wilson, Harvard University, Cambridge, MA
Robert Wilson, Petronal, New York, NY
Ronald Wilson, Hopkins Uni-Physics Lab, Laural, MD
Anne Wittenberg, ICF, Inc., Washington, DC
Eliza Wojtaszek, Battelle Memorial Institute, Washington, DC
John Wozniak, Hopkins Uni-Physics Lab, Laurel, MD
Bunli Yang, Ontario Ministry of Energy, Toronto, Ontario, CD
Terry Yonker, Great Lakes United, Buffalo, NY

INDEX

GPSR Compliance

The European Union's (EU) General Product Safety Regulation (GPSR) is a set of rules that requires consumer products to be safe and our obligations to ensure this.

If you have any concerns about our products, you can contact us on ProductSafety@springernature.com

In case Publisher is established outside the EU, the EU authorized representative is:

Springer Nature Customer Service Center GmbH
Europaplatz 3
69115 Heidelberg, Germany

Batch number: 10761599

Printed by Printforce, the Netherlands